MAN AND THE ENVIRONMENT INFORMATION GUIDE SERIES

Series Editor: Seymour M. Gold, Assistant Professor, Department of Environmental Horticulture, University of California, Davis, California

Also in this series:

AIR POLLUTION—*Edited by George Hagevik* **

ENVIRONMENTAL DESIGN—*Edited by Wolfgang F. E. Preiser and Steven Parshall* **

ENVIRONMENTAL ECONOMICS—*Edited by Warren E. Johnston* **

ENVIRONMENTAL EDUCATION—*Edited by William B. Stapp and Mary Dawn Liston*

ENVIRONMENTAL LAW—*Edited by Mortimer D. Schwartz* *

ENVIRONMENTAL PLANNING—*Edited by Michael J. Meshenberg* *

ENVIRONMENTAL POLITICS AND ADMINISTRATION—*Edited by Nedjelko D. Suljak* **

ENVIRONMENTAL TOXICOLOGY—*Edited by Robert Rudd* **

ENVIRONMENTAL VALUES, 1860-1972—*Edited by Loren C. Owings* *

HUMAN ECOLOGY—*Edited by Gary H. Winkel and Emilie O'Mara* **

NOISE POLLUTION—*Edited by Clifford R. Bragdon* *

WATER POLLUTION—*Edited by Allen W. Knight and Mary Ann Simmons* **

*in press
**in preparation

The above series is part of the
GALE INFORMATION GUIDE LIBRARY

The Library consists of a number of separate Series of guides covering major areas in the social sciences, humanities, and current affairs.

General Editor: Paul Wasserman, Professor and former Dean, School of Library and Information Services, University of Maryland

WASTEWATER MANAGEMENT

WASTEWATER MANAGEMENT

A GUIDE TO INFORMATION SOURCES

*Volume 2 in the Man and the Environment
Information Guide Series*

George Tchobanoglous

*Associate Professor
University of California, Davis*

Robert Smith

*Project Engineer
Metcalf & Eddy*

Ronald Crites

*Project Engineer
Metcalf & Eddy*

Gale Research Company
Book Tower, Detroit, Michigan 48226

**Library of Congress
Cataloging in Publication Data**

Tchobanoglous, George.
   Wastewater management.

   (Man and the environment information guide series;
v. 2) (Gale information guide library)
   Includes indexes.
     1. Sewage disposal–Bibliography. I. Crites,
Ronald W., joint author. II. Smith, Robert Gordon,
1947-   joint author. III. Title.
Z5853.S22T34 [TD741]   016.3636   74-11570
ISBN 0-8103-1338-3

VITAE

George Tchobanoglous is currently an associate professor of environmental engineering in the Department of Civil Engineering at the University of California at Davis. His research interests include wastewater and industrial waste treatment, solid waste management, small treatment systems, and water quality in aquatic systems.

Tchobanoglous received his B.S. from the University of Pacific, his M.S. from the University of California at Berkeley, and his Ph.D. from Stanford University. He has authored or coauthored over fifty articles, papers, and reports dealing with various environmental engineering topics. He is a registered engineer in California and Montana.

Robert G. Smith is currently pursuing a Ph.D. degree in environment engineering at the University of California at Davis. From 1970 to 1975 he served as a project engineer with Metcalf & Eddy, Engineers, in Palo Alto, California.

Smith received his B.S. in civil engineering and his M.S. in environmental engineering from Stanford University. He was a contributor to WASTEWATER ENGINEERING and is a registered engineer in California.

Ronald W. Crites is currently a project engineer with Metcalf & Eddy, Engineers, in Palo Alto, California. In his six years with M & E he has specialized in wastewater reclamation and land treatment of wastewater.

Crites received his B.S. in civil engineering from Chico State College and an M.S. and Engineer's degree from Stanford University. He has authored or coauthored eight articles and papers on wastewater treatment and has contributed to WASTEWATER ENGINEERING and LAND TREATMENT AND DISPOSAL OF MUNICIPAL AND INDUSTRIAL WASTEWATER. He is a registered engineer in California.

CONTENTS

Contents

FOREWORD

The rapidly expanding scope and sophistication of literature in the area of environmental studies has created the need for an authoritative, annotated guide to the range and quality of information sources. New legislation, publications, developing techniques, and increasing levels of public involvement in environmental issues have created a need for government, industry, organizations, and individuals to seek more authoritative information on man-environment relationships. This information can be used to rationalize proposed public or private actions, resolve issues, and measure the impact of past actions on man and the environment.

The objective of this series is to provide an authoritative and systematic guide to significant information sources on selected topics for use by scholars, students, scientists, reference librarians, consultants, professionals, and citizens. The scope and scale of each volume are directed toward accommodating both the immediate and continuing needs of users in addition to establishing a chronological benchmark of the literature on each topic.

This series is concerned with the cause and effect relationships of man's impact on the urban and natural environment. It emphasizes the human problems, issues, and implications of these relationships that can be used to study or solve environmental problems. The series is intended to be a primary reference for people not familiar with the location and organization of information sources in the area of environmental studies. It should also serve as a basic reference for advanced students or professionals who need an annotated and selected sample of the current literature.

Each volume is prepared by a noted authority on the topic. The annotations and selection of references have been carefully reviewed to present an objective and balanced array of thought and practice. The organization of each volume is tailored to emphasize the way information is organized in this multidisciplinary and rapidly evolving field.

This volume is an authoritative and comprehensive reference work on wastewater management that brings together a diverse set of information sources from the physical, social, and natural sciences.

Foreword

The volume presents a compendium of reference material and ideas invaluable to the scholar and practitioner. It contains a highly selected collection of periodical literature, government documents, scientific journals, and teaching materials on the rapidly evolving aspect of environmental engineering. The introductions, annotations, and sources have been carefully written to give the reader a technical and pragmatic grasp of problems involved in wastewater management.

The scholarly nature of annotations, classification, and cross-referencing of materials presents an integrated view of time, space, and events that is not common to other reference works in this area. The authors have made a very technical subject accessible to practitioners and students with their clear and concise organization of a wide range of sources.

As the second volume in this series, WASTEWATER MANAGEMENT continues the standard of scholarly excellence and problem solving utility that will be followed in eleven additional volumes now in preparation. When the entire series is completed in 1976, it will represent the most authoritative and comprehensive guide to the man-environmental literature ever published.

Seymour M. Gold, Series Editor
November 1975

PREFACE

This guide is intended to be a selective rather than an exhaustive work. It is designed to consolidate the literature in the field of wastewater management into a pertinent and useful information source and is intended for use by anyone who is initiating a literature search in this field. For purposes of this guide, wastewater management has been defined to encompass the engineering of wastewater collection, treatment, disposal, and reuse systems; the economic analysis and planning of such systems; and the legislation affecting the implementation of these systems.

In terms of organization, the aforementioned subject areas have been divided into seven separate chapters. An introductory chapter is included to provide perspective on the subject area by defining basic needs or unresolved issues and suggesting possible solutions or directions toward solutions. Throughout the book, only those references that were felt to be most useful to anyone who is initiating a literature search were included. In particular, references were selected that contained either an exceptionally clear and concise coverage of specific topics or consolidated the literature.

The number of annotated references under a specific topic generally has been limited to four or less to keep this volume to a useful size. In some areas of wastewater management, where rapid development is occurring and literature consolidation has not yet been accomplished, it was not possible to cover the subject adequately with this many references. In these cases, additional references have been included without annotations to provide representative coverage. These additional references are considered to be less useful than those that are annotated, but may cover specific topics not found elsewhere.

Most of the material included in the eight chapters that comprise the main body of the guide are either periodical articles, technical reports, or collected papers. General reference works and texts are included in Appendix A. The types of works cited are limited to those dealing principally with the theory and fundamentals of various aspects of wastewater management. The intent is to present works that will not soon be outdated as a result of developments in specific areas. However, if a specific topic is covered particularly well in a text or reference work, that section of the book is cited and annotated separately under the specific topic in the main body of the guide. Periodicals and journals, newsletters, and abstracts and digests related to wastewater management are listed and annotated separately in Appendices B, C, and D. A list of names and addresses of federal and state governmental agencies involved in water pollution control is presented in Appendix E and a list of publishers and their addresses in Appendix F.

To aid the user of this guide in identifying and retrieving the references, each of the annotated references in the main body of the guide and those in Appendix A are number-coded by Section (e.g., 204 refers to the fourth reference listed in Section 2, a03 refers to the third entry in Appendix A). These number codes, rather than page numbers, are used for retrieval in the author, title, and subject indexes.

LIST OF ABBREVIATIONS

The following abbreviations for journal and periodical titles, organizations, and governmental agencies have been used in this guide.

JOURNALS AND PERIODICALS

JOUR. AWWA - JOURNAL OF THE AMERICAN WATER WORKS ASSOCIATION

JOUR. ENV. ENGR. DIV. ASCE - JOURNAL OF THE ENVIRONMENTAL ENGINEERING DIVISION OF THE AMERICAN SOCIETY OF CIVIL ENGINEERS

JOUR. SAN. ENGR. DIV. ASCE - JOURNAL OF THE SANITARY ENGINEERING DIVISION OF THE AMERICAN SOCIETY OF CIVIL ENGINEERS

JOUR. WPCF - JOURNAL OF THE WATER POLLUTION CONTROL FEDERATION

AGENCIES AND ORGANIZATIONS

ASCE - American Society of Civil Engineers

EPA - U.S. Environmental Protection Agency

FWPCA - Federal Water Pollution Control Administration

FWQA - Federal Water Quality Administration

SCS - U.S. Soil Conservation Service

SERL - Sanitary Engineering Research Laboratory, University of California

WPCF - Water Pollution Control Federation

Section 1

PROBLEMS, NEEDS, SOLUTIONS: A PERSPECTIVE

Section 1

PROBLEMS, NEEDS, SOLUTIONS: A PERSPECTIVE

An overview of wasterwater management is presented in this section. Questions of what needs to be done and how solutions can be developed and implemented are addressed. Trends in collection and treatment followed by disposal of effluents and sludges are identified.

Specific topics included in this section deal with (1) the goals and objectives of wastewater management, (2) wastewater management systems, (3) economic and financial considerations, (4) planning and implementation, and (5) legislation and legal control. It should be noted that the information sources abstracted in this section are general in scope and coverage as compared to the more specific sources presented in the subsequent sections.

A. GOALS AND OBJECTIVES

The consequences of inadequate wastewater management may include risks to public health and the degradation of the terrestrial and aquatic environment. What constitutes pollution, and where and to what extent it should be controlled, are among the crucial issues that must be addressed.

Documentation

*   Need for wastewater treatment
*   Effects of discharging untreated wastewater

Unresolved Issues

*   What constitutes pollution
*   Degree of treatment required
*   Whether stream or effluent standards should be used
*   How to best achieve pollution control

101 Christman, R.F., et al. THE NATURAL ENVIRONMENT: WASTES

3

AND CONTROL. Pacific Palisades, Calif.: Goodyear Publishing Co., 1973. 228 p.

The extent of scientific knowledge and the strengths and deficiencies of some major theories of environmental phenomena are discussed. Processes and concepts are emphasized with the intention of introducing the reader to the problems of environmental pollution. Many pollution control devices are described in a concise manner without bogging the reader down in the technicalities. The overall focus of the book is on what is known about air and water pollution and solid waste management. For anyone who is not an expert in the field and does not understand the current knowledge regarding environmental issues, this book will be of value.

102 Henderson, J.M. "Water Pollution - Facts and Fantasies." JOUR. SAN. ENGR. DIV. ASCE 98(June 1972): 529-46.

The author notes that the immediate goal of present federal action in water pollution control is to require secondary treatment as a minimum regardless of stream use and quality, waste and stream flows, or ecological needs. Many streams used as fisheries require the addition of artificial fertilizer for optimum levels of aquatic life. The fact that migration of populations from rural to urban areas is neglected in establishing blanket pollution control regulations is stressed. The pollution problems requiring the greatest attention are related mainly to the 75 percent of the population located in 2 percent of the land area.

B. WASTEWATER MANAGEMENT SYSTEMS

Wastewater management systems are composed of the facilities required for the collection, treatment, and disposal of wastewater including sludge residues. The ultimate objective in the design, construction, and operation of wastewater management systems is to eliminate water pollution at an affordable cost.

Documentation

*   Basic theory of unit operations and processes
*   Flowsheets to achieve specific effluent levels

Unresolved Issues

*   Optimum combination of processes and operations
*   Effects of constituents in treated wastewater on water reuse
*   Effects of alternative disposal methods

General

103   Okun, D.A.   "New Directions for Wastewater Collection and Disposal."
      JOUR. WPCF 43(November 1971): 2171-80.

      A general discussion of current trends in wastewater management
      is presented. Topics covered include: regional organization,
      wastewater reuse, wastewater characteristics, wastewater treat-
      ment, and manpower needs. The discussion is necessarily brief,
      but the important issues are highlighted.

Collection

104   Rogers, B.R.   "Sewers - Hidden But Vital."   JOUR. WPCF 42(October
      1971): 2170.

      This is a short editorial identifying the need to place more em-
      phasis on sewer systems in grant programs and research efforts.
      Perspective on the role of sewers in wastewater management is
      provided.

Treatment

105   Barth, E.F.   "Perspectives on Wastewater Treatment Processes: Physical-
      Chemical and Biological."   JOUR. WPCF 43(November 1971): 2189-94.

      The author identifies the areas in complete physical-chemical
      treatment schemes that require more investigation before this
      treatment approach can come into common use as an alterna-
      tive to biological treatment schemes. It is concluded that in
      the near future municipal wastewater treatment technology will
      center on biological systems, with specific physical-chemical
      processes to meet specific requirements.

106   Thomas, R.E., and Harlin, C.C., Jr.   "Experiences with Land Spreading
      of Municipal Effluents."   Presented at the First Annual IFAS Workshop on
      Land Renovation of Wastewater, June 1972, Tampa, Florida.   13 p.

      The three general categories of land application (1) infiltration-
      percolation, (2) cropland irrigation, and (3) spray-runoff are
      described.   Infiltration-percolation for groundwater recharge can
      be used on coarse textured soils with high loadings under ideal
      conditions.   Crop irrigation by spraying or spreading requires
      more land than infiltration-percolation, as loadings are general-
      ly less than eight feet per year.   Spray-runoff is especially
      suited to impermeable soils since treatment occurs on the surface
      and more than half the applied effluent is returned directly to
      surface waters as planned and controlled runoff.   EPA-sponsored
      research activities in the field in California, Minnesota, Penn-
      sylvania, and Michigan are summarized.

Sludge Handling and Disposal

107   Burd, R.S. A STUDY OF SLUDGE HANDLING AND DISPOSAL.
FWPCA Water Pollution Control Research Series, no. WP-20-4. Cincin-
nati: FWPCA, May 1968. 369 p. Paperbound. Sold by Government
Printing Office.

Various methods of sludge handling and disposal with emphasis
on theoretical development, application examples, and economics are documented. The material is presented in the same sequence as solids are processed in treatment plants, beginning
with the grit chamber and ending with ultimate sludge disposal.
In all, some twenty operations and processes are discussed separately, with a detailed reference list for each subject. A
summary of relative costs for various methods is given. Some
ideas for future approaches to separate the solid portion of
wastewater and to treat it more effectively are included. (Also
cited in Section 5, A.)

C. ECONOMIC CONSIDERATIONS

Cost effectiveness in wastewater management involves the comparison of alternative systems on a common basis in which program costs and benefits have
been quantified. Development of basic cost data, assessment of alternatives,
and governmental grant programs are specific economic considerations in such
an analysis.

Documentation

*   Cost of treatment facilities
*   Conventional techniques in cost-benefit analysis
*   Procedures for valuation and economic assessment

Unresolved Issues

*   Basis for equitable user charges for sewerage services
*   Federal or local payment of operating costs
*   Interest rates and return periods for municipal works

a09   Environmental Protection Agency. GUIDANCE FOR FACILITIES PLAN-
NING. 2nd ed. Washington, D.C.: 1974. 115 p. Paperbound.

This guide is a supplement to federal regulations on grants for
construction of publicly owned treatment works. The concept
of cost-effectiveness analysis required to be used in the planning of such facilities is defined. In a separate chapter devoted to the subject of monetary cost analysis, the elements to
be considered in an analysis are defined and the method of

analysis is described. Useful examples of analysis are presented. (For general review, see Appendix A.) (Also cited in Section D and Section 6, C.)

108 McJunkin, F.E. COSTS OF WATER POLLUTION CONTROL, PROCEEDINGS OF A NATIONAL SYMPOSIUM. Raleigh: North Carolina State University, April 1972. 276 p. Illustrations. Paperbound.

The needs and issues related to the economics of water pollution control are discussed by the thirty-three authors and speakers. The individual papers are followed by a summary of questions and answers from the audience. The viewpoints of industry, federal and state agencies, university professors, consulting engineers, and environmentalists are expressed with emphasis on cost effectiveness as a concept.

D. PLANNING AND IMPLEMENTATION

Environmental planning involves determination of the quality and diversity of existing ecosystems and the development of alternative strategies to minimize any adverse environmental impacts associated with waste disposal. In this connection, wastewater management planning has become increasingly important as expanding populations exert greater impact on the biosphere. Although regional management is becoming more popular, implementation problems are numerous.

Documentation

*   Federal guidelines
*   Generalized prediction techniques

Unresolved Issues

*   Validity of regionalization
*   Definition of environmental quality

109 Bella, D.A., and Overton, W.S. "Environmental Planning and Ecological Possibilities." JOUR. SAN. ENGR. DIV. ASCE 98(June 1972): 579-92.

Man's inability to predict the outcome of his activities on the biosphere is identified as the cause of the current environmental predicament. In the future, it is reasoned, man's ability to influence and modify the environment will increase faster than his ability to foresee the effects of his actions. An environmental strategy is developed, based on the avoidance of large-scale, irreversible change and the preservation of environmental diversity. It is concluded that such a strategy has the potential of reducing the chances of unanticipated conditions detrimental to the quality of life.

110 Environmental Protection Agency. WATER QUALITY STRATEGY PAPER, A STATEMENT OF POLICY FOR IMPLEMENTING THE REQUIREMENTS OF THE 1972 FEDERAL WATER POLLUTION CONTROL ACT AMENDMENTS. Washington, D.C.: 15 March 1974. 82 p. Paperbound. Sold by Government Printing Office.

The near- and long-term objectives of the EPA in implementing the 1972 Federal Water Pollution Control Act amendments are identified. Final policy is not presented, but rather the best thinking at the time of writing. Designed to be revised annually, the paper aims to set forth the activities and priorities of each forthcoming year. This paper presents a good overview of planning on a national level.

a09 Environmental Protection Agency. GUIDANCE FOR FACILITIES PLANNING. 2nd ed. Washington, D.C.: 1974. 115 p. Paperbound.

This guide is a supplement to federal regulations on grants for construction of publicly owned treatment works and is intended for the use of those involved in the planning of such facilities. A delineation of the steps required in the planning process and a detailed listing of federal regulations and pertinent references are presented. The relationships of facilities plans to basin plans, area-wide waste treatment management plans, and municipal permits are explained. A sample outline of a facilities plan is included. The guide will be updated periodically to incorporate changes and additional information. (For general review, see Appendix A.) (Also cited in Section 1, C and Section 6, C.)

E. LEGISLATION AND LEGAL CONTROL

Only a few principal federal and state laws form the basis for the legal control of water pollution in the United States. Regulations issued by various governmental agencies, as well as court decisions, serve as a basis for the interpretation of these laws.

Documentation

*   Public laws

8

Unresolved Issues

*   Uniformity in the application of laws
*   Interpretation of legislation in specific instances

111   Alberi, L.T.   "Environmental Law - The Litigation Controversy."   JOUR.
      WPCF 43(December 1971): 2463-66.

      In this article, litigation in environmental protection is con-
      sidered mainly to be a forum for general public involvement in
      environmental issues.   The usefulness of litigation in solving
      pollution problems is considered from a number of viewpoints.

112   Reitze, A.W.   ENVIRONMENTAL LAW.   2nd ed.   Washington, D.C.:
      North American International, 1972.   813 p.

      This book covers laws and litigation related to general envi-
      ronmental policy, air pollution, solid wastes, and water pol-
      lution.   In the chapter on water pollution, considerable back-
      ground material is presented to familiarize the reader with some
      of the technical aspects of the problem.   The text of principal
      federal legislation is presented along with commentary.   State
      and local water pollution control are discussed briefly.   Several
      important cases involving water pollution control are reviewed.

Section 2

WASTEWATER COLLECTION

Section 2

WASTEWATER COLLECTION

The subject areas considered in this section include: (1) wastewater sources and quantities; (2) sewer planning, design, and construction; (3) urban stormwater management; (4) sewer maintenance and operation; and (5) sewer use ordinances. General references on wastewater collection are included at the beginning of this section followed by more specific information sources.

Until recently, there was little research and development in the field of wastewater collection. Much of the current research activity has been stimulated by the need to reduce or eliminate infiltration and inflow into sewer systems and the recognition of stormwater as a significant pollution source.

It should be noted that the subject area of stormwater management encompasses much more than collection. For this reason, no attempt has been made to review individually all the many aspects of stormwater management in this section. Instead, comprehensive references that cite more specific references are presented.

A. GENERAL

a01   Babbitt, H.E., and Baumann, R.E. SEWERAGE AND SEWAGE TREAT-
       MENT. 8th ed. New York: John Wiley & Sons, 1958. Pp. 10-322.

       The first half of this text is devoted to sewage collection sys-
       tems. Planning, sewage flows, hydraulics, design, construc-
       tion, pumping, and maintenance are discussed. Some of the
       information is dated, particularly that on construction materials
       and equipment. (For general review, see Appendix A.) (Also
       cited in Section 3, C.)

a21   Metcalf & Eddy, Inc. WASTEWATER ENGINEERING: COLLECTION,
       TREATMENT, DISPOSAL. San Francisco: McGraw-Hill, 1972. Pp.
       13-225.

       Chapter 2 through 6 are devoted to wastewater collection.
       Separate chapters are devoted to the determination of waste-
       water flow rates, the principles of sewer hydraulics, the de-

sign of sewers, sewer appurtenances and special structures, and pumps and pumping stations. The chapter on pumping stations is one of the most complete references on the subject. (For general review, see Appendix A.) (Also cited in Section 3, A, B, C, E, F, and H; Section 4, B; and Section 5, A.)

B. WASTEWATER SOURCES AND QUANTITIES

The term "sources" as used in wastewater sources is assumed to refer to the various specific origins of stormwater and domestic sewage. Flow character-istics such as quantity of flow, variations in flow rate, and peak stormwater flows are considered under the heading "wastewater quantities." Prediction of stormwater flows and pollution loads by computer modeling is an important recent development.

Documentation

*   Representative per capita flow rates
*   General quantities and sources
*   Effects of flow reduction at sources

Unresolved Issues

*   Effects of system interconnections
*   Contributions from illicit connections, such as roof drains
*   Infiltration control methodology

201   Friedland, A.O.; Shea, T.G.; and Ludwig, H.F. "Quantity and Qual-ity Relationships For Combined Sewer Overflows." In ADVANCES IN WATER POLLUTION RESEARCH: PROCEEDINGS OF THE 5TH INTERNA-TIONAL CONFERENCE, pp. I-1/1 to I-1/16. Edited by S.H. Jenkins. New York: Pergamon Press, 1971.

The results of a one-year program of wet and dry weather mon-itoring of several combined sewer systems and one storm sewer system in San Francisco are described. The quantity and qual-ity of combined sewage flows from differing land use types are compared, and the relative contributions of dry weather and storm runoff flows to combined sewage flows are determined. Pollution abatement by treatment and by storage is also evalu-ated. Although this is a case study, the conclusions reached are generally applicable to most urban areas.

202   Schmidt, J.O. "Pollution Control in Sewers." JOUR. WPCF 44(July 1972): 1384-92.

The extent of extraneous flows into separate sanitary sewers is described on a nationwide basis using selective survey results.

The common sources of infiltration and inflow into sanitary sewers are identified and discussed. The major sources identified include basement drains and house service sewers. Corrective measures designed to minimize extraneous flows in existing and new sewers are described briefly through examples. A concise review of the subject of infiltration-inflow is given.

C. SEWER PLANNING, DESIGN, AND CONSTRUCTION

Urban planning, growth prediction, and the general layout of collection systems to meet future needs are encompassed in sewer planning. Sewer design includes the functional layout of sewers and the selection of specific sizes and slopes for sewers to provide needed capacity. Construction considerations include sewer pipe materials and techniques of pipe installation.

Documentation

*   Specific design procedures
*   Design and form of standard appurtenances
*   Hydraulics of gravity sewers
*   Construction materials

Unresolved Issues

*   Design alternatives for the control of anaerobic growths
*   Design and use of pressure and vacuum systems
*   Optimization of sewer system design using computer techniques

Gravity Sewers

203   Argaman, Y.; Shamir, U.; and Spivak, E. "Design of Optimal Sewerage Systems." JOUR. ENV. ENGR. DIV. ASCE 99(October 1973): 703-16.

Describes a procedure by which the least expensive sewerage systems may be designed by using dynamic programming techniques. The presentation is technical and presents program formulations. Practical limitations of the method are discussed, and a good definition of the state of the art of optimal sewer design using computer techniques is provided. At present, the procedure may be used to suboptimize small networks that may be combined to form large networks.

204   Joint Committee of American Society of Civil Engineers and Water Pollution Control Federation. DESIGN AND CONSTRUCTION OF SANITARY AND STORM SEWERS. Manual of Practice no. 9. Washington, D.C.: Water Pollution Control Federation, 1970. 331 p.

Designed as an aid to the practicing engineer, this manual
covers all aspects of sewerage projects from inception to com-
pletion of construction. Topics covered include: investigations
and surveys, quantities of sewage and stormwater, sewer system
design including appurtenances and special structures, construction materials and procedures, pumping stations, and contract
documents. A thorough reference on the subject.

205 Paintal, A.S. "Hydraulic Design of Self-Cleaning Circular Sanitary
Sewers." WATER & SEWAGE WORKS 119(1972): R-52.

Describes a method for selecting sewer size and slope based on
the development of a critical boundary shear stress that will
initiate sewage solids movement. The basic mathematical relationships necessary to determine the principal variables are
derived. Useful design charts incorporating these relationships
are presented for several conditions. The method of design is
offered as an alternative to the method of minimum velocity
commonly used in sewer design.

206 Pomeroy, R.D. "Sanitary Sewer Design for Hydrogen Sulfide Control."
PUBLIC WORKS 101(October 1970): 93.

The conditions within a sewer line that result in sulfide generation and build-up are reviewed, and formulations are presented
for predicting the occurrence of sulfide build-up. It is noted
that actual sewer failures due to sulfide-related corrosion of
concrete pipe are not common. Chemical control measures to
inhibit sulfide production are described. Design considerations
to avoid sulfide production, such as minimum slopes and velocities, are reviewed. "Rule of thumb" numbers are presented.

Pressure Sewers

207 American Society of Civil Engineers. COMBINED SEWER SEPARATION
USING PRESSURE SEWERS. FWPCA Water Pollution Control Research
Series, ORD-4. Washington, D.C.: October 1969. 198 p. Paper-
bound.

A feasibility report on the use of pressure sewers to separate
combined sewer systems. A review of information on the practical application of the concept is presented. The proposed
system would involve storage-grinder pumps for each residential
or commercial sanitary sewer connection. The equipment to be
used is described as well as the installation of pressure conduits
in existing systems. The relative cost and reliability of pressure systems versus gravity systems are evaluated. A thorough
and objective reference on the subject.

208 Carcich, I.G., et al. "The Pressure Sewer: A New Alternative to
Gravity Sewers." CIVIL ENGINEERING - ASCE 44(May 1974): 50-53.

A pressure sewer system involves a grinder pump placed in the basement of the dwelling with a small-diameter outlet pipe to the forcemain. The advantages of such systems under special conditions such as rocky terrain and waterside developments are cited. The possibilities are extensive, and the reader will surely ponder other uses for pressure systems. An EPA-sponsored demonstration project in Albany, New York, is described, and some evaluation of potential problems is given.

D. URBAN STORMWATER MANAGEMENT

Stormwater management comprises the collection, regulation, and treatment of stormwater. Stormwater collected in combined sewer systems is of primary concern. The methods of treatment covered under wastewater treatment in Section 3 are often used for stormwater treatment.

Documentation

* Pollution loads contributed by stormwater
* Techniques for stormwater pollution abatement

Unresolved Issues

* Effects of polymers in increasing sewer capacity
* Optimization of stormwater pollution control

209 Field, R., and Struzeski, E.J., Jr. "Management and Control of Combined Sewer Overflows." JOUR. WPCF 44(July 1972): 1393-1415.

An overview of the federal government's involvement in the determination and abatement of combined sewer overflow pollution is contained in this paper. The background and extent of combined sewer overflow problems are reviewed. Existing and newly developing control and treatment systems for overflow management are described. A good concise review of the state of the art of the subject, and a good reference source, as more than 130 references are cited.

210 Lager, J.A., and Smith, W.G. URBAN STORMWATER MANAGEMENT AND TECHNOLOGY: AN ASSESSMENT. EPA Contract no. 68-03-0179. Washington, D.C.: EPA, May 1974. 446 p. Paperbound.

The results of an investigation and assessment of the state of the art of urban stormwater management are presented and discussed in this comprehensive report. Management alternatives discussed include treatment and control as well as the collection of stormwater. A wealth of information is presented on current stormwater control technology in a textbook format. The doc-

umentation is extensive. The conclusions and recommendations
are helpful in providing the reader with some perspective on
the general area of stormwater management.

E. SEWER OPERATION AND MAINTENANCE

Sewer maintenance and operation includes routine cleaning, inspection, and
small repairs, as well as the major rehabilitation of collection systems. To
retain the full flow capacity of the collection system and to minimize infiltration, a regular inspection and maintenance program is necessary.

Documentation

* Use of television cameras for inspection
* Sewer cleaning techniques
* Use of plastic linings for sewer rehabilitation

Unresolved Issues

* Sulfide and odor control techniques
* Sewer and storm sewer cleaning frequency
* The effectiveness of catch basins

211 Bremner, R.M. "In-place Lining of Small Sewers." JOUR. WPCF
43(July 1971): 1444-56.

Upgrading existing sewers by lining with flexible polyethylene
pipe is described in this paper. Methods used to insert the
lining are illustrated with photos and drawings. Interesting
before-and-after photos are also included. Although the sewer
system in Toronto, Canada, is referred to specifically, the article is quite informative regarding the general application of
this rapidly developing method of sewer rehabilitation.

212 Santry, I.W., Jr. "Sewer Maintenance Costs." JOUR. WPCF 44(July
1972): 1425-32.

Typical data required for a sewer maintenance program are presented along with summaries of maintenance costs for the Dallas,
Texas, collection system in 1969-70. The relative costs of labor, materials, and equipment for various maintenance activities
are presented. Although this is a case study, the information
presented is generally applicable.

213 Water Pollution Control Federation, Subcommittee on Sewer Maintenance.
SEWER MAINTENANCE. Manual of Practice no. 7. Washington, D.C.:
1966. 58 p. Paperbound.

This manual is intended to be a working reference for sewer maintenance supervisory personnel and staff as well as design engineers. Methods of organizing maintenance operations and keeping records, as well as commonly used maintenance equipment, are described. Preventive sewer maintenance procedures are stressed. Design considerations to aid in sewer maintenance are reviewed. In addition to being a concise review of the subject, it includes numerous additional references.

214   The Western Co. POLYMERS FOR SEWER FLOW CONTROL. FWPCA Water Pollution Control Research Series, no. WP-20-22. Washington, D.C.: FWPCA, August 1969. 180 p. Paperbound.

Several different polymers were investigated to determine the effect of their injection into sewers on the flow capacity of the sewers and their effect on aquatic flora and fauna. It was demonstrated with laboratory and field tests that significant flow increases can be achieved with polymer injection. Economic analyses are presented, and it is shown that this practice can be considerably less expensive than new construction in preventing sewer overflows during peak storm flows. Much field testing is still required.

F.   SEWER USE ORDINANCES

Sewer ordinances are public laws that are used to regulate the installation of sewers and the subsequent discharge of wastes into these sewers. The types of wastewater that may be discharged into sewers, considering public safety and compatability with treatment processes, are usually specified in most ordinances.

Documentation

*   Ordinances for construction and installation of sewers

Unresolved Issues

*   Constituents to be excluded
*   Use of ordinances to control population growth
*   Industrial pretreatment requirements

215   Water Pollution Control Federation, Subcommittee on Regulation of Sewer Use. "Manual of Practice no. 3: Regulation of Sewer Use." JOUR. WPCF 45 (September 1973): 1985-2012 and (October 1973): 2216-35.

A draft of a revised manual of practice on regulation of sewer use is presented in two consecutive issues to allow Journal readers the opportunity to comment on the proposed revisions. The proposed draft is an update of the current manual.

Wastewater Collection

216   Water Pollution Control Federation, Subcommittee on Municipal Sewer
Ordinances. REGULATION OF SEWER USE. Manual of Practice no. 3.
Washington, D.C.: 1968. 41 p. Paperbound.

The component parts of a recommended sewer use ordinance are
delineated in detail. A suggested ordinance is provided to illustrate the application of the various components. Several additional references on the subject are provided, but most are
pre-1960. A good introduction to the subject.

Section 3

WASTEWATER TREATMENT

Section 3

WASTEWATER TREATMENT

Wastewater treatment must be considered the focal point of wastewater management. The importance of this subject area is reflected in the relative number of information sources cited in this section. Although this section is more extensive than the others, the references selected are, for the most part, those that consolidate essential information and serve as a good source for additional references. In rapidly developing areas of interest like advanced wastewater treatment, such summarizing references have not yet evolved. Consequently, more specific references describing original research are cited in these areas. In some cases additional references are cited without annotations to provide representative coverage.

The development of new and improved treatment processes has been and is receiving considerable federal support through research grants. These research efforts serve as the principal source of new information in the field. Current emphasis is on advanced wastewater treatment and upgrading existing facilities. Construction of treatment works is being accelerated by the federal construction grant program established under the Federal Water Quality Control Act Amendments of 1972.

A. WASTEWATER CHARACTERISTICS

Wastewater characteristics refer to the physical, chemical, and biological properties of wastewater that affect treatment processes or are taken as a pollutive measure of the strength or impact of the wastewater.

Documentation

*   Representative characteristics of municipal wastewater
*   Typical incremental addition of salts
*   Techniques for constituent analysis
*   Techniques for assessing treatability

Wastewater Treatment

Unresolved Issues

* Definition of the sources and variability of trace elements
* Seasonal variability of characteristics
* Chemical interactions between constituents during collection

301 DeFilippi, J.A., and Shih, C.S. "Characteristics of Separated Storm and Combined Sewer Flows." JOUR. WPCF 43(October 1971): 2033-58.

The results of an extensive monitoring program designed to determine the characteristics of wastewater collected in combined sewers and storm sewers are presented. The extensive amount of data collected are well presented, allowing a good comparison to be made between dry- and wet-weather flow conditions. The conclusions of the study provide insight into the nature of storm flows (quantity and quality) and clearly demonstrate the significant pollution load associated with storm flows.

a11 Fair, G.M.; Geyer, J.C.; and Okun, D.A. WATER AND WASTEWATER ENGINEERING. Vol. 2. New York: John Wiley & Sons, 1968. Pp. 20-15 to 20-17.

A brief discussion of the average composition of municipal wastewaters and per capita contributions is included. (For general review, see Appendix A.) (Also cited in Section 3, B, E, and H.)

302 Ligman, K.; Hutzler, N.; and Boyle, W. "Household Wastewater Characterization." JOUR. ENV. ENGR. DIV. ASCE 100(February 1974): 201-13.

The results of a household wastewater generation and sampling survey are presented. A small number of rural and urban households were surveyed to determine the frequency of specific wastewater discharge events and the characteristics of the individual discharges. Constituents analyzed included 5-day biochemical oxygen demand (BOD$_5$), total solids, suspended solids, total dissolved solids, and volatile solids. Although not very scientific in approach, this study is one of the few reported attempts to characterize individual household wastewater.

a21 Metcalf & Eddy, Inc. WASTEWATER ENGINEERING: COLLECTION, TREATMENT, DISPOSAL. San Francisco: McGraw-Hill, 1972. Pp. 227-71.

One chapter is devoted to a comprehensive classification and description of wastewater constituents. Measurement of wastewater constituents is discussed, and a few of the more important analytical techniques are presented. (For general review, see Appendix A.) (Also cited in Section 2, A; Section 3, B, C, E, F, and H; Section 4, B; and Section 5, A.)

a23   Nemerow, N.L.   LIQUID WASTE OF INDUSTRY: THEORIES, PRAC-
TICES, AND TREATMENT. Reading, Mass.: Addison-Wesley Publishing
Co., 1971. 584 p.

The author categorizes major wastewater-generating industries
and describes the origin and general character of the various
wastewaters. It is impossible, of course, to accurately de-
scribe all industrial wastewater characteristics. To solve a
given industrial wastewater management problem, specific data
must be obtained for the wastewater in question. The very
general characterizations presented in this book can be helpful as a starting point in focusing attention on the major
constituents when dealing with an unknown wastewater. (For
general review, see Appendix A.)

303   Schwinn, D. E., and Dickson, B.H.   "Nitrogen and Phosphorus Variation
in Domestic Wastewater." JOUR. WPCF 44(November 1972): 2059.

The variation of nitrogen and phosphorus in domestic wastewater
is evaluated for flows entering three different treatment plants.
The extensive amount of data collected was analyzed statisti-
cally with the aid of computer programs designed for the purpose. The correlation of nitrogen and phosphorus concentration
with flow, biochemical oxygen demand (BOD), and suspended
solids was established along with the peaking characteristics of
the nutrient concentrations. Although this was a case study,
the conclusions drawn are significant from a design standpoint.
The nitrogen and phosphorus concentrations were concluded to
be independent of flow, BOD, or suspended solids concentration.

304   Zanoni, A.E., and Rutkowski, R.J.   "Per Capita Loadings of Domestic
Wastewater." JOUR. WPCF 44(September 1972): 1756-62.

This survey represents one of the few recent attempts to characterize domestic sewage on a per capita basis. Sewage flow
from a strictly residential area with a population of 1,207 was
measured and sampled over a six-month period. The results
obtained are significant because many of the observed values
are substantially lower than those commonly used, but based on
much earlier work. Per capita, 5-day biochemical oxygen demand (BOD5) loadings were found to be 0.10 lb/capita/day,
while suspended solids loadings were 0.08 lb/capita/day.

B. THEORY AND FUNDAMENTALS OF TREATMENT

The basic concepts, principles, and theories applied to the analysis and design
of the various unit operations and processes used for wastewater treatment are
collectively referred to as the theory and fundamentals of treatment.

Wastewater Treatment

Documentation

* Fundamental principles of physical unit operations
* Process and operational kinetics for conventional treatment processes

Unresolved Issues

* Mechanisms responsible for removal of specific constituents
* Optimum sequential application of unit operations and processes

General

all Fair, G.M.; Geyer, J.C.; and Okun, D.A. WATER AND WASTEWATER
ENGINEERING. Vol. 2. New York: John Wiley & Sons, 1968.
668 p.

> The theory of the basic unit operations of water and wastewater
> treatment is reviewed. Chapters dealing with the fundamentals
> of water chemistry and aquatic biology are included. The use-
> fulness of this text as a practical working reference is limited.
> (For general review, see Appendix A.) (Also cited in Section
> 3, A, E, and H.)

a21 Metcalf & Eddy, Inc. WASTEWATER ENGINEERING: COLLECTION,
TREATMENT, DISPOSAL. San Francisco: McGraw-Hill, 1972. Pp.
273-421.

> Three chapters are devoted to theory and fundamentals of phys-
> ical unit operations, chemical unit processes, and biological
> unit processes. The basic principles are highlighted, but not
> oversimplified. The material is intended primarily for students
> and practicing engineers. (For general review, see Appendix
> A.) (Also cited in Section 2, A and Section 3, A.)

Physical Unit Operations

a32 Weber, W.J., Jr. PHYSICOCHEMICAL PROCESSES FOR WATER QUAL-
ITY CONTROL. New York: John Wiley & Sons, 1972. 666 p.

> This is a comprehensive text on both theory and application of
> physical unit operations and chemical unit processes employed
> in wastewater treatment. It is useful for students and practic-
> ing engineers because the theory presented is not abstract, but
> related to practical application. (For general review, see Ap-
> pendix A.) (Also cited in Section 3, B below, E, F, and H
> and Section 5, A.)

305 McCabe, W.L., and Smith, J.C. UNIT OPERATIONS OF CHEMICAL

ENGINEERING. New York: McGraw-Hill, 1967. 1007 p.

This is a textbook designed for the undergraduate student of chemical engineering. Most of the principles presented, however, are applicable to the unit operations used in wastewater treatment. The major topics covered include: (1) fluid mechanics, (2) heat transfer, (3) mass transfer, and (4) operations involving particulate solids. It is a good introductory text, especially for chemical engineers.

a25 Rich, L.G. UNIT OPERATIONS OF SANITARY ENGINEERING. New York: John Wiley & Sons, 1961. 308 p:

Fundamentals of physical unit operations encountered in sanitary and chemical engineering are delineated. The various operations are considered apart from particular applications. The material presented is based on the chemical engineering approach and notation.

Biology and Biological Unit Processes

306 Lawrence, A.W., and McCarty, P.L. "A Unified Basis for Biological Treatment Design and Operation." JOUR. SAN. ENGR. DIV. ASCE 96(June 1970): 757-78.

This paper contains one of the clearest presentations on the design of biological systems using basic kinetic equations. The mean cell residence time or solids retention time is suggested as the most useful parameter both in design and control of biological systems. Mathematical models for complete-mix systems and plug-flow systems are developed and the applications of the models to various forms of biological processes are discussed. A complete reference list on the theory of biological kinetics is also included.

a17 McKinney, R.E. MICROBIOLOGY FOR SANITARY ENGINEERS. New York: McGraw-Hill, 1962. 293 p.

Written for sanitary engineering students, the fundamentals of microbiology and the biochemistry of microorganisms are presented on an introductory level. Emphasis is placed on the microbiology of mixed cultures in dilute nutrient substrates. The second half of the book is devoted to applied microbiology in which the fundamentals are related to the major biological unit processes: (1) trickling filters, (2) activated sludge, (3) oxidation ponds, and (4) anaerobic digestion. Although some of the coverage is dated, this is still a useful reference. (For general review, see Appendix A.)

Wastewater Treatment

Chemistry and Chemical Unit Processes

a29 Sawyer, C.N., and McCarty, P.L. CHEMISTRY FOR SANITARY EN-
GINEERS. 2nd ed. San Francisco: McGraw-Hill, 1967. Pp. 1-285.

> An excellent introductory textbook in chemistry for the sanitary
> engineer. Major topics presented include basic concepts from
> (1) qualitative chemistry, (2) quantitative chemistry, (3) organic
> chemistry, (4) physical chemistry, (5) colloid chemistry, (6)
> biochemistry, and (7) radiochemistry. It also serves as a good
> basic reference for those interested in more thorough coverage
> of specific topics. (For general review, see Appendix A.)
> (Also cited in Section 3, H and J.)

a32 Weber, W.J., Jr. PHYSICOCHEMICAL PROCESSES FOR WATER QUAL-
ITY CONTROL. New York: John Wiley & Sons, 1972. 666 p.

> A useful teaching text on both theory and application of phys-
> ical unit operations and chemical unit processes employed in
> wastewater treatment. (For general review, see Appendix A.)
> (Also cited in Section 3, B above, E, F, and H and Section 5, A.)

C. CONVENTIONAL PHYSICAL TREATMENT OPERATIONS

Conventional physical treatment operations are those in which treatment is
achieved by the application of physical forces. They are usually associated
with preliminary and primary treatment. The operations considered include:
screening, grit removal, and sedimentation.

Documentation

*   Basic theory
*   Expected performance
*   General application and design parameters

Unresolved Issues

*   Alternative applications of operations
*   Effect of energy input on flocculation
*   Performance prediction of aerated grit chamber and screens

General

a21 Metcalf & Eddy, Inc. WASTEWATER ENGINEERING: COLLECTION,
TREATMENT, DISPOSAL. San Francisco: McGraw-Hill, 1972. Pp. 423-44.

> This text contains a good basic description, including design
> parameters, for conventional physical treatment operations

normally referred to as preliminary and primary treatment. Helpful illustrations and examples of equipment layouts for screening, grit removal, and sedimentation facilities are included. (For general review, see Appendix A.) (Also cited in Section 2, A; Section 3, A, B, C below, E, F, and H; Section 4, B; and Section 5, A.)

307 Joint Committee of the Water Pollution Control Federation and American Society of Civil Engineers. SEWAGE TREATMENT PLANT DESIGN. Manual of Practice no. 8. Washington, D.C.: 1972. 375 p. Paperbound.

Theoretical and practical aspects of design of conventional physical operations are presented. The theory presented is quite basic and emphasis is placed on practical considerations. Chapters are included on flotation and flocculation as well as screening, grit removal, and sedimentation. Conventional biological treatment and sludge treatment and handling processes are also covered in the manual, although updating is needed for these topics. (Also cited in Section 3, H.)

Screening

a19 Metcalf, L., and Eddy, H.P. AMERICAN SEWERAGE PRACTICE. Vol. 3: DISPOSAL OF SEWAGE. 3rd ed. New York: McGraw-Hill, 1935. 909 p.

A separate chapter of this text is devoted to the topic of racks and screens. Although much of the equipment described is outdated, the design considerations discussed are still valid. Information presented on the quantity and character of screenings for various screen sizes is quite useful. (See also below.)

Grit Removal

a01 Babbitt, H.E., and Baumann, R.E. SEWERAGE AND SEWAGE TREATMENT. 8th ed. New York: John Wiley & Sons, 1958. Pp. 420-30.

Theoretical and practical aspects involved in the design of a constant velocity grit chamber are presented. Velocity control devices such as proportional weirs and measuring flumes are described. A useful design example for a constant-velocity grit chamber and control flume is presented. (For general review, see Appendix A.) (Also cited in Section 2, A and Section 3, C below.)

308 Neighbor, J.B., and Cooper, T.W. "Design and Operation Criteria for Aerated Grit Chambers." WATER & SEWAGE WORKS 112(December 1965): 448-54.

Design criteria for aerated grit chambers are reviewed, including air supply rates, detention times, and circulation velocities. The presentation is somewhat general with regard to tank sizing. The findings of a survey of grit chambers' performance are presented, but no clear-cut recommended design changes are indicated. The effects of baffling on grit removal efficiency are documented with test results.

Sedimentation

309   Camp, T.R. "Sedimentation and the Design of Settling Tanks." TRANS-ACTIONS ASCE 111(1946): 895-958.

One of the first papers in which the known principles of sedimentation were organized and presented in a form that allowed for the development of a design theory that can be used in practice. Several design examples employing design theory are presented. The importance of overflow rates as opposed to detention times as a design parameter is stressed. Insights contained in the discussions which follow the paper are as valuable as the paper itself.

310   Dick, R.I., and Ewing, B.B. "Evaluation of Activated Sludge Thickening Theories." JOUR. SAN. ENGR. DIV. ASCE 93(August 1967): 9-29.

Although attention is focused on activated sludge separation, the fundamental considerations presented are applicable to all sedimentation operations in which thickening or compaction occurs. The development of thickening theories is traced through a brief literature review. The results of experimental studies are used to support the significant conclusion that the settling rate of flocculant sludge is dependent upon sludge depth and mixing of underlying layers as well as sludge concentration.

D. CONVENTIONAL BIOLOGICAL TREATMENT PROCESSES

Conventional biological treatment is achieved by bacterial assimilation of organic material. Such processes are commonly used for secondary treatment and are often referred to as secondary treatment processes. The processes considered include Imhoff and septic tanks, activated sludge, trickling filters, biodisks, stabilization ponds, and aerated lagoons.

Documentation

*   Process kinetics including temperature effects
*   Physical configuration of processes

* General design criteria
* Qualitative descriptions
* General applications and expected performance

Unresolved Issues

* Prediction of effluent quality from sedimentation tanks
* Factors controlling reliability
* Effects of variations in wastewater characteristics on process performance
* Process control

General

311 Austin, Texas. City of, and Center for Research in Water Resources. DESIGN GUIDES FOR BIOLOGICAL WASTEWATER TREATMENT PROCESSES. EPA Water Pollution Control Research Series, 11010 ESQ. Washington, D.C.: EPA, August 1971. 221 p. Paperbound. Sold by Government Printing Office.

An excellent general reference for the major types of biological processes. Much of the text deals with basic kinetic relationships in biological processes. Design using theoretical equations is emphasized with examples presented for each type of process. Operating data from several plants are used in discussing practical application of design equations. A useful section on wastewater characteristics, quantity and quality, is included. The importance of considering variations in flow and composition of wastewater as part of design is stressed.

a21 Metcalf & Eddy, Inc. WASTEWATER ENGINEERING: COLLECTION, TREATMENT, DISPOSAL. San Francisco: McGraw-Hill, 1972. Pp. 481-573.

A separate chapter is devoted to the design of biological facilities. Theoretical and practical aspects of design receive equal attention. A useful reference. (For general review, see Appendix A.) (Also cited in Section 2, A; Section 3, A, B, C above, E, F, and H; Section 4, B; and Section 5, A.)

Septic Tanks and Imhoff Tanks

a01 Babbitt, H.E., and Baumann, R.E. SEWERAGE AND SEWAGE TREATMENT. 8th ed. New York: John Wiley & Sons, 1958. Pp. 467-80.

One chapter is devoted to a discussion of septic tanks and Imhoff tanks. The section on septic tanks is limited, with tank design being the only subject covered. Leaching fields are not mentioned. The section on Imhoff tanks is quite descriptive and complete. Specific design criteria are presented.

(For general review, see Appendix A.) (Also cited in Section 2, A and Section 3, C above.)

a19   Metcalf, L., and Eddy, H.P. AMERICAN SEWERAGE PRACTICE. Vol. 3: DISPOSAL OF SEWAGE. 3rd ed. New York: McGraw-Hill, 1935. 909 p.

The basics of design and operation of both septic tanks and Imhoff tanks are presented in detail in this classic text. Examples of existing systems are used extensively in discussing design and process control techniques. Entertaining from the historical standpoint. (See also above.)

312   U.S. Public Health Service. MANUAL OF SEPTIC TANK PRACTICE. PHS Publication no. 526. Washington, D.C.: 1967. 92 p. Paperbound. Sold by Government Printing Office.

This manual consists of two parts: (1) septic tanks - soil adsorption systems for private residences and (2) systems for institutions and recreational areas. It serves as a practical guide to the design and construction of septic tank systems. Soil adsorption systems, trenches, seepage pits, and distribution arrangements are discussed, as well as septic tank design and specifications. Procedures for percolation tests are also described.

Activated Sludge

313   Burchett, M.E., and Tchobanoglous, G. "Facilities for Controlling the Activated Sludge Process by Mean Cell Residence Time." JOUR. WPCF 46(May 1974): 973-79.

The basic methods used by operators to control the activated sludge process are discussed in this article. The fundamental mathematical expressions describing biological growth are presented to explain the theoretical basis for control methods. The mean cell residence time (MCRT) is demonstrated to be the most suitable control parameter. A practical system that allows control of the MCRT is described in detail. This is a clear, concise explanation of both the theoretical and practical aspects of process control methods.

314   Dick, R.I. "Role of Activated Sludge Final Settling Tanks." JOUR. SAN. ENGR. DIV. ASCE 96(April 1970): 423-36.

The need to consider the final settling tank more thoroughly in designing activated sludge plants is stressed. An example using actual operating data is presented to show how failure to consider the thickening function of final settling tanks can lead to unsatisfactory performance of the entire activated sludge process. The fundamental basis for rational design of the final settling tanks using settling test data are described.

315　McKinney, R.E., and O'Brien, W.J. "Activated Sludge: Basic Design Concepts." JOUR. WPCF 40(November 1968): 1831-43.

A brief history of the development of the activated sludge processes is presented in which major contributions are cited. The major modifications to the activated sludge process, step aeration, contact stabilization, and complete-mix are described. The importance of providing adequate mixing through design is emphasized. Criteria are presented for primary design variables, including feed pattern, detention time, microbe concentration, aeration equipment and requirements, and tank configuration. A good, general review of the process with regard to practical design considerations.

316　Water Pollution Control Federation, Subcommittee on Aeration in Wastewater Treatment. AERATION IN WASTEWATER TREATMENT. Manual of Practice no. 5. Washington, D.C.: 1971. 90 p. Paperbound.

Although, strictly speaking, aeration is a physical unit operation, it is an integral part of most aerobic biological processes, particularly activated sludge. This manual is designed to be a practical working reference for designers and operators. The various applications of aeration in wastewater treatment are reviewed. The determination of oxygen requirements for biological process is discussed. The types and applications of diffused aeration equipment and mechanical aeration equipment are described in detail, and manufacturers are referenced. A section is included on operation and maintenance of aeration equipment. Diffused aeration equipment appurtenances, blowers, air piping, and air filters are described.

317　Wilcox, E.A., and Thomas, A. "Oxygen Activated Sludge Wastewater Treatment Systems." EPA Technology Transfer Seminar Publication. Washington, D.C.: EPA, August 1973. 46 p. Paperbound.

This publication contains a review of operating experience with the Unox pure oxygen process including the results of government-sponsored studies and independent demonstrations. Process design procedures are presented along with criteria for independent design variables. The presentation of information strongly emphasizes the advantages of the oxygen process, hardly surprising since the principal author is with Unox.

Trickling Filters

318　Dow Chemical Co. A LITERATURE SEARCH AND CRITICAL ANALYSIS OF BIOLOGICAL TRICKLING FILTER STUDIES. 2 vols. EPA Water Pollution Control Research Series, 17050 DDY. Washington, D.C.: EPA, December 1971. Vol. 1, 341 p.; vol. 2, 361 p. Paperbound.

Sold by Government Printing Office.

A complete review of nearly all information sources on the subject of trickling filters up to the publication date may be found in these two volumes. A critique of the literature cited follows each subtopic and provides a much-needed perspective on the more than 5,000 references cited. Conclusions derived from the literature review are useful in defining the current status of the trickling filter applications and identifying the critical parameters that should be considered when using trickling filters.

Rotating Biological Disks

319 Antonie, R.L.; Kluge, D.L.; and Mielke, J.H. "Evaluation of a Rotating Disk Wastewater Treatment Plant." JOUR. WPCF 46(March 1974): 498-511.

A 0.5 million gallons per day (mgd) rotating disk biological treatment plant at Pewaukee, Wisconsin, was evaluated over a nine-month period under an EPA grant. Variables investigated included hydraulic loadings and exposure to varying climatic conditions. Results are not correlated to the primary design factor, detention time, but BOD and ammonia removals are related to plant flow. Stable operation is achieved through winter and summer conditions. Power consumption is also calculated by the authors.

320 Torpey, W.N., et al. "Rotating Disks with Biological Growths Prepare Wastewater for Disposal or Reuse." JOUR. WPCF 43(November 1971): 2181-88.

Discusses a pilot plant system in which rotating disks with biological growths were used to provide various degrees of organic and nutrient removal. Initial stages of disks provided removal of carbonaceous material with succeeding stages accomplishing nitrification. The unique feature of this system was a final series of illuminated disks on which algae were generated and nitrogen and phosphorous removal achieved. Results of this part of the study are incomplete. Insufficient lighting for algae growth is noted.

Aerated Lagoons

321 Bartch, E.H., and Randall, C.W. "Aerated Lagoons - A Report on the State of the Art." JOUR. WPCF 43(April 1971): 699-708.

The important distinction between aerobic or complete-mix aerated lagoons and facultative aerated lagoons is stressed. Most aerated lagoons fall into the facultative category. This type of lagoon is the main topic of this paper. A brief but clear explanation of why the performance of facultative lagoons

cannot be accurately predicted from present kinetic theory is
presented. The sensitivity of lagoon efficiency to temperature,
a disadvantage in cold climates, is demonstrated with operational data.

322   Missouri Basin Engineering Health Council. WASTE TREATMENT LA-
GOONS: STATE OF THE ART. EPA Water Pollution Control Research
Series, 17090 EHX. Washington, D.C.: EPA, June 1971. 152 p.
Paperbound. Sold by Government Printing Office.

A section of this report is devoted to aerated lagoons. An interesting and incisive history of aerated lagoon development is
presented which provides a good overview. Several different
industrial waste applications are cited. The development of
aeration equipment for lagoon applications is traced although
propeller-type floating aerators are not mentioned. Kinetic
theory of complete-mix lagoons is presented. Facultative lagoons are also discussed. (See also below.)

323   Sawyer, C.N. "New Concepts in Aerated Lagoon Design and Operation." In ADVANCES IN WATER QUALITY IMPROVEMENT, pp. 325-
35. Edited by E.F. Gloyna and W.W. Eckenfelder, Jr. Austin: University of Texas Press, 1968.

Practical aspects of aerated lagoon application are presented
and discussed. Emphasis is placed on temperature considerations
because lagoons are excellent heat dissipaters. Treatment efficiency in lagoons is sensitive to temperature changes, particularly for short detention times. The advantage of aerated lagoons
in terms of operating flexibility is identified. Aeration equipment selection is discussed briefly. Useful for design.

Stabilization Ponds

324   Barsom, G. LAGOON PERFORMANCE AND THE STATE OF LAGOON
TECHNOLOGY. Environmental Protection Technology Series, EPA-R2-
73-144. Washington, D.C.: EPA, June 1973. 214 p. Paperbound.
Sold by Government Printing Office.

A critical review of stabilization pond design and operating
practices is presented, based on a survey of municipal systems.
The causes of common pond failures, such as odors, short-cir-
cuiting, and algae carry-over, are identified; and recommendations are presented to avoid such failures through proper design
and operation. Hardly a literary gem, but the report is candid
and makes several valid points regarding pond design and operation in the conclusions and recommendations.

325   Gloyna, E.F. WASTE STABILIZATION PONDS. Geneva, Switzerland:
World Health Organization, 1971. 175 p.

A general reference written for the design engineer which presents and discusses the basic theory of biological treatment processes, design criteria for ponds, alternative design approaches, and common operational problems. A unique feature of the book is a worldwide survey of stabilization pond use with a brief description of practices within various countries.

326 Missouri Basin Engineering Health Council. WASTE TREATMENT LAGOONS: STATE OF THE ART. EPA Water Pollution Control Research Series, 17090 EHX. Washington, D.C.: EPA, July 1971. 152 p. Paperbound. Sold by Government Printing Office.

A summary of existing information on design and operation of all types of lagoon treatment systems including stabilization ponds, aerated lagoons, and anaerobic ponds is presented. Future needs to improve pond systems are identified. It is concluded that pond effluent quality will not always meet current water quality criteria and that poor quality is associated with microbial solids in the effluent. Although some basic theory is presented, this reference will be of use primarily to those interested in practical considerations. (See also above.)

327 Oswald, W.J. "Advances in Anaerobic Pond System Design." In ADVANCES IN WATER QUALITY IMPROVEMENT, pp. 409-26. Edited by E.F. Gloyna and W.W. Eckenfelder, Jr. Austin: University of Texas Press, 1968.

The importance of optimizing anaerobic digestion in facultative and anaerobic ponds is discussed with reference to several cases. Pond depth is the primary design variable that can be used to control environmental conditions within a pond system. It is concluded that the most efficient and nuisance-free system is an anaerobic-facultative-aerobic series of ponds with recirculation from aerobic to anaerobic. An excellent summary of practical considerations for pond design is included. The volume in which this paper appears has an entire section devoted to waste stabilization pond practice with other noted authors' on theory, current practice, and the treatment of industrial waste in ponds.

Stabilization Ponds—Additional References

327a Brown and Caldwell Consulting Engineers. UPGRADING LAGOONS. EPA Technology Transfer Seminar Publication. Washington, D.C.: EPA, August 1973. 43 p. Paperbound.

327b McKinney, R.E., ed. SECOND INTERNATIONAL SYMPOSIUM FOR WASTE TREATMENT LAGOONS. Lawrence: University of Kansas, June 1970. 404 p. Paperbound.

E. PHYSICAL-CHEMICAL TREATMENT PROCESSES

Unit operations and processes in which treatment is achieved through the combined application of physical forces and chemical reactions are often referred to as physical-chemical treatment processes. Although biological reactions are not included theoretically, they often take place simultaneously. The ability to treat toxic wastewaters and handle shock loadings are the principal advantages of physical-chemical processes.

Documentation

* General application
* Physical configuration
* Generalized design criteria

Unresolved Issues

* Optimum chemical dosages and points of application
* Optimum application
* Sludge handling and disposal
* Removal of specific constituents

General

all Fair, G.M.; Geyer, J.C.; and Okun, D.A. WATER AND WASTEWATER ENGINEERING. Vol. 2. New York: John Wiley & Sons, 1968. Pp. 24-1 to 31-29.

A review of fundamentals and design considerations for physical-chemical processes. Comprehensive, but somewhat theoretical. (For general review, see Appendix A.) (Also cited in Section 3, A, B, and H.)

a12 Gloyna, E.F., and Eckenfelder, W.W., Jr., eds. WATER QUALITY IMPROVEMENT BY PHYSICAL AND CHEMICAL PROCESSES. Austin: University of Texas Press, 1970. 448 p. (Published for the Center for Research in Water Resources.)

A collection of papers by several noted authors on the functional design considerations of physical-chemical treatment processes. Processes considered include filtration, sedimentation, coagulation, precipitation, carbon adsorption, and ion exchange. In general, these papers deal with specific aspects of the particular processes and generally do not provide an overall review. (For general review, see Appendix A.)

a21 Metcalf & Eddy, Inc. WASTEWATER ENGINEERING: COLLECTION, TREATMENT, DISPOSAL. San Francisco: McGraw-Hill, 1972. Pp. 423-80.

Wastewater Treatment

A general discussion of applied physical-chemical treatment, including various flow sheets. Design aspects are emphasized. (For general review, see Appendix A.) (Also cited in Section 2, A; Section 3, A, B, C, F, and H; Section 4, B; and Section 5, A.)

a32 Weber, W.J., Jr. PHYSICOCHEMICAL PROCESSES FOR WATER QUALITY CONTROL. New York: John Wiley & Sons, 1972. Pp. 61-408.

A text devoted to fundamentals of physical-chemical processes. Application and design aspects are discussed for both water and wastewater treatment. (For general review, see Appendix A.) Also cited in Section 3, B, F, and H and Section 5, A.)

Complete Treatment

328 Bishop, D.F.; O'Farrell, T.P.; and Stamberg, J.B. "Physical-Chemical Treatment of Municipal Wastewater." JOUR. WPCF 44(March 1972): 361-71.

A concise summary of recent studies on physical-chemical treatment is presented, and the development of technology in this area is traced. The results of a one hundred gallons per minute (gpm) pilot study conducted at Washington, D.C. are presented. The physical-chemical process consisted of two-stage lime precipitation, filtration, selection ion exchange for ammonia removal, and granular carbon adsorption. Several uncertainties associated with physical-chemical treatment are apparent but not specifically identified in this paper.

329 Burns, D.E., and Shell, G.L. PHYSICAL-CHEMICAL TREATMENT OF A MUNICIPAL WASTEWATER USING POWDERED CARBON. Environmental Protection Technology Series, EPA-R2-73-264. Washington, D.C.: EPA, August 1973. 230 p. Paperbound. Sold by Government Printing Office.

Physical-chemical treatment of municipal wastewater was studied on a one hundred gpm pilot scale. The process consisted of chemical coagulation-precipitation using solids contactors followed by powdered carbon contacting and granular media filtration. Biological activity in the carbon contactors enhanced organic removals. Powdered carbon regeneration was only partially successful. A good background is presented on each of the processes employed. The discussion of results and development of design criteria is informative and worthwhile.

330 Shuckrow, A.J.; Gaynor, D.W.; and Bonner, W.F. PHYSICAL-CHEMICAL TREATMENT OF COMBINED AND MUNICIPAL SEWAGE. Environmental Protection Technology Series, EPA-R2-73-149. Washington,

D.C.: EPA, February 1973. 178 p. Paperbound. Sold by Government Printing Office.

A physical-chemical process for raw wastewater treatment was developed on a laboratory scale and demonstrated with 0.1 mgd pilot plant. The process consists of contacting with powdered carbon, coagulating with alum and polyelectrolytes, settling, and filtering with tri-media filters. The combined carbon and alum sludge is dewatered and passed through a fluidized-bed furnace for carbon regeneration. Alum is recovered by acidifying. A secondary quality effluent was obtained with the wastewater studied. The study is significant because it demonstrated successfully the efficient regeneration of powdered carbon.

331 Weber, W.J., Jr.; Hopkins, C.B.; and Bloom, R.S. "Physicochemical Treatment of Wastewater." JOUR. WPCF 42(January 1970): 83-99.

This paper contains the results of one of the first of the recent pilot studies on physical-chemical treatment. The process studied consisted of coagulation-precipitation using ferric chloride, followed by dual media filtration and granular carbon adsorption using both packed and expended beds. It was concluded that coagulation-precipitation and carbon adsorption could provide the degree of treatment needed for most applications, and that filtration should be used at the end of the system as a polishing step in applications requiring exceptionally clear effluents. As with many similar papers, objectivity is lacking. The authors emphasize the attributes of physical-chemical treatment without adequately identifying drawbacks or unknowns.

Physical-Chemical Processes—Additional References

331a Burns, D.E., and Shell, G.L. "Carbon Treatment of a Municipal Wastewater." JOUR. WPCF 46(January 1974): 148-64.

331b CH2M/Hill, Engineers. PHYSICAL-CHEMICAL WASTEWATER TREATMENT PLANT DESIGN. EPA Technology Transfer Seminar Publication. Washington, D.C.: EPA, August 1973. 41 p. Paperbound.

331c Hager, D.G., and Reilly, P.B. "Clarification-Adsorption in the Treatment of Municipal and Industrial Wastewater." JOUR. WPCF 42(May 1972): 794-800.

331d Libby, D.V., et al. "Chemical-Physical Treatment of Municipal Wastewater Using Continuous Countercurrent Filtration." JOUR. WPCF 44(April 1972): 574-82.

331e Rizzo, J.L., and Schade, R.E. "Secondary Treatment with Granular

Activated Carbon." WATER & SEWAGE WORKS 116(August 1969): 307-12.

331f Zuckerman, M.M., and Molof, A.H. "High Quality Reuse Water by Chemical-Physical Wastewater Treatment." JOUR. WPCF 42(March 1970): 437-63.

F. ADVANCED WASTEWATER TREATMENT PROCESSES

The term "advanced wastewater treatment" is used to define those processes or treatment systems that provide treatment beyond that achievable with secondary treatment. Such treatment processes may follow or work in conjunction with conventional secondary treatment schemes. Advanced wastewater treatment may involve removal of any or all of the following constituents: nutrients, suspended solids, dissolved organics, and dissolved solids. Removal of specific industrial constituents is not considered.

Documentation

* Qualitative description
* Generalized design criteria
* General application and performance data

Unresolved Issues

* Factors affecting process reliability
* Optimum application in process flowsheets
* Process selection for removal of nitrogen, dissolved organics, and minerals

General

a06 Culp, R.L., and Culp, G.L. ADVANCED WASTEWATER TREATMENT. New York: Van Nostrand Reinhold Co., 1971. 310 p.

This is the first book to be devoted to the subject of full-scale application of advanced wastewater treatment (AWT). The discussion of the various processes is well balanced between basic principles, design considerations, and operating performance. The breadth of coverage of treatment processes, however, is quite limited. Only those processes employed in the South Lake Tahoe reclamation plant receive in-depth coverage. One chapter, devoted to other methods, presents a brief description of other types of processes, including ion exchange, ozonation, reverse osmosis, nitrification-denitrification, powdered carbon, combined biological-chemical treatment and physical-chemical treatment. A selected list of references is presented under each process category. (For general review, see Appendix A.)

332 Environmental Protection Agency. PROCEEDINGS: ADVANCED WASTE
TREATMENT AND WATER REUSE SYMPOSIUM, DALLAS, TEXAS. 5 vols.
Washington, D.C.: January 1971. Paperbound.

This collection of papers is presented in five separate volumes,
each dealing with a different aspect of advanced wastewater
treatment. Due to the rapidly changing state of the art of
AWT, some of the information on the status of research is
slightly dated. These papers, however, contain the most complete general discussion of AWT available in one reference.
Papers in addition to those presented at the symposium have
been included. The volume on nitrogen removal in particular
offers complete general coverage.

a21 Metcalf & Eddy, Inc. WASTEWATER ENGINEERING: COLLECTION,
TREATMENT, DISPOSAL. San Francisco: McGraw-Hill, 1972. Pp.
.633-72.

A general review of available processes including application
and performance. A brief discussion of basic principles for
several processes, including ammonia stripping, filtration, ion
exchange, and nitrification-denitrification. (For general review, see Appendix A.) (Also cited in Section 2, A; Section 3,
A, B, C, E, and H; Section 4, B; and Section 5, A.)

333 University of California, Sanitary Engineering Research Laboratory.
WASTEWATER RECLAMATION AND REUSE WORKSHOP PROCEEDINGS.
Berkeley: June 1970. 313 p. Paperbound.

This is a collection of papers by several noted authorities covering nearly all facets of advanced wastewater treatment. Although somewhat out of date with regard to the status of research efforts, excellent summaries of AWT processes, fundamentals, and performance are presented. Of particular interest
are papers on nitrogen, phosphate, and mineral removal. Following each paper is a discussion that provides, in most cases,
an objective interpretation.

Nitrogen Removal

334 Kepple, L.G. "Ammonia Removal and Recovery Becomes Feasible."
WATER & SEWAGE WORKS 121(April 1974): 42.

Describes a process in which ammonia released from an air stripping unit is absorbed in an acid solution and recovered as an
ammonium salt. The stripping unit is operated in the conventional manner except that the gas stream is recycled rather than
using outside air. Such recycling eliminates the major obstacles
of scaling and freezing. The process can be used to treat the

main stream or spent regenerant from ion exchange processes. If the process proves to be reliable on a full scale, it could be a major breakthrough in nitrogen removal.

335   Lance, J.C.   "Nitrogen Removal by Soil Mechanisms." JOUR. WPCF 44(July 1972): 1352-61.

The removal mechanisms and transformations involved with nitrogen in the soil are discussed in detail. Land disposal removes nitrogen from wastewater by denitrification, volatilization of ammonia, and removal of plant tissue grown on the site. Denitrification is perhaps the most desirable removal process because nitrogen is transferred to the atmosphere as gaseous nitrogen without causing air pollution. Irrigating grasses with secondary effluent should not result in nitrogen pollution of the groundwater, providing the nitrogen applied does not greatly exceed the nitrogen taken up by the growing crop.

336   McCarty, P.L. "Biological Processes for Nitrogen Removal: Theory and Application." In PROCEEDINGS, TWELFTH SANITARY ENGINEERING CONFERENCE, pp. 136-52. Edited by V. Snoeyink: Urbana: University of Illinois, Department of Civil Engineering, 1970. Paperbound.

Nitrogen removal by bacterial assimilation, algae harvesting, and nitrification-denitrification is discussed in depth. A clear and concise review of the removal mechanisms involved in each process is presented. The limitations in applying each removal process are discussed through reference to first-hand experience and the recent literature. Based on cost, reliability, and potential removal efficiency, nitrification-denitrification appears to be the most promising of the biological nitrogen removal processes.

337   Reeves, T.G. "Nitrogen Removal: A Literature Review." JOUR. WPCF 44(October 1972): 1895-1908.

Attention in this review is focused on nitrogen removal by air stripping, ion exchange, and biological nitrification-denitrification. A brief description of the theory of each process is followed by a review of literature. Although some of the literature reviewed is not current, the conflicting viewpoints regarding applicability of the processes are presented and discussed. Other processes, including electrochemical treatment, demineralization, breakpoint chlorination, algae harvesting, and land application, are described briefly with very little review.

338   Tchobanoglous, G. "Physical and Chemical Processes for Nitrogen Removal: Theory and Application." In PROCEEDINGS, TWELFTH SANITARY ENGINEERING CONFERENCE, pp. 110-36. Edited by V. Snoeyink. Urbana: University of Illinois, Department of Civil Engineering, 1970. Paperbound.

A good brief review of the basics of physical and chemical nitrogen removal processes. Covered in depth are air stripping, ion exchange, and chlorination. Several other less developed processes are discussed briefly. Useful cost data referenced to the Engineering News Record Construction Cost Index are presented on each of the processes.

Nitrogen Removal—Additional References

338a Atkins, P.F., et al. "Ammonia Removal by Physical-Chemical Treatment." JOUR. WPCF 45(November 1973): 2372-88.

338b Haug, R.T., and McCarty, P.L. "Nitrification with Submerged Filters." JOUR. WPCF 44(November 1972): 2086-2102.

338c Huang, C., and Hopson, N.E. "Nitrification Rate in Biological Processes." JOUR. ENV. ENGR. DIV. ASCE 100(April 1974): 409-22.

338d Mercer, B.W., et al. "Ammonia Removal from Secondary Effluent by Selective Ion Exchange." JOUR. WPCF 42(February 1970): R95-107.

338e Metcalf & Eddy, Inc. NITRIFICATION AND DENITRIFICATION FACILITIES. EPA Technology Transfer Seminar Publication. Washington, D.C.: EPA, August 1973. 36 p. Paperbound.

338f O'Farrell, T.P., et al. "Nitrogen Removal by Ammonia Stripping." JOUR. WPCF 44(August 1972): 1527-35.

338g Requa, D.A., and Shroeder, E.D. "Kinetics of Packed-Bed Denitrification." JOUR. WPCF 45(August 1973): 1696-1707.

338h Stensel, H.D.; Loehr, R.C.; and Lawrence, A.W. "Biological Kinetics of Suspended-Growth Denitrification." JOUR. WPCF 45(February 1973): 249-61.

338i University of California, Sanitary Engineering Research Laboratory. OPTIMIZATION OF AMMONIA REMOVAL BY ION EXCHANGE USING CLINOPTILOLITE. EPA Water Pollution Control Research Series, 18080 DAR. Washington, D.C.: EPA, September 1971. 189 p. Paperbound. Sold by Government Printing Office.

338j Wuhrmann, K. "Objectives, Technology, and Results of Nitrogen and Phosphorous Removal Processes." In ADVANCES IN WATER QUALITY IMPROVEMENT, pp. 21-44. Edited by E.F. Gloyna and W.W. Eckenfelder, Jr. Austin: University of Texas Press, 1969.

Phosphorus Removal

339 Black & Veatch, Consulting Engineers. PROCESS DESIGN MANUAL
FOR PHOSPHORUS REMOVAL. EPA Program no. 17010 GNP. Washington, D.C.: EPA, October 1971. 191 p.

This manual is limited in its coverage to the removal of phosphorus by chemical precipitation using iron and aluminum salts
and lime. Chemical addition at several possible points in a
treatment system is covered. Points of addition include (1) before primary clarifiers, (2) in trickling filters, (3) in activated
sludge aeration basins, (4) before secondary clarifiers, and (5)
in tertiary processes. A description of each treatment method
is presented along with performance data, equipment requirements, design parameters, and cost. The manual is useful to
those interested in design, operation, and performance of chemical treatment systems.

340 Minton, G.R., and Carlson, D.A. "Combined Biological-Chemical
Phosphorus Removal." JOUR. WPCF 44(September 1972): 1736-55.

A clear and very thorough review of phosphorus removal by
dosing iron and aluminum salts directly to biological systems is
presented and supported by an extensive literature review. A
summary of performance for phosphorus removal in the primary
clarifier and in tertiary processes is also presented. The importance of the dosage point in the chemical-biological system
is demonstrated through a review of precipitation mechanisms
and reported experimental results. It is concluded that alum
and ferric ion should be dosed as close to the final clarifier as
possible, while sodium aluminate should be dosed as close as
possible to the aeration basin outlet.

341 Shindala, A. "Nitrogen and Phosphorus Removal from Wastewaters: Part
1 and Part 2." WATER & SEWAGE WORKS 119(June 1972): 66-71 and
(July 1972): 60-67.

A concise review of proven and potential phosphorus removal
processes is presented in this two-part paper. Operative removal mechanisms in the various processes and control of process
variables to optimize performance are the primary topics of discussion. The types of phosphorus removal processes considered
include biological-chemical precipitation, chemical-biological,
chemical-physical, and ion exchange. Documentation through
literature citation is quite good. A similar, though less extensive, review of nitrogen removal by nitrification-denitrification,
air stripping, and ion exchange is also presented.

Phosphorus Removal—Additional References

341a Bell, G.R.; Libby, D.V.; and Lordi, D.T. PHOSPHORUS REMOVAL

USING CHEMICAL COAGULATION AND A CONTINUOUS COUNTER-CURRENT FILTRATION PROCESS. FWQA Water Pollution Control Research Series, 17010 EDO. Washington, D.C.: EPA, June 1970. 57 p. Paperbound. Sold by Government Printing Office.

341b Carberry, J.B., and Tenney, M.W. "Luxury Uptake of Phosphate by Activated Sludge." JOUR. WPCF 45(December 1973): 2444-62.

341c Ferguson, J.F., and McCarty, P.L. THE PRECIPITATION OF PHOS-PHATES FROM FRESH WATER AND WASTEWATERS. Department of Civil Engineering, Stanford University, Technical Report no. 120. Stanford, Calif.: Department of Civil Engineering, Stanford University, December 1969. 202 p. Paperbound.

341d Kumar, J.I., and Clesceri, N.L. "Phosphorous Removal from Waste-waters: A Cost Analysis." WATER & SEWAGE WORKS 120(March 1973): 82-91.

341e Leckie, J.O., and Stumm, W. "Phosphate Precipitation." In WATER QUALITY IMPROVEMENT BY PHYSICAL AND CHEMICAL PROCESSES, pp. 237-49. Edited by E.F. Gloyna and W.W. Eckenfelder, Jr. Austin: University of Texas Press, 1970.

341f Menar, A.B., and Jenkins, D. CALCIUM PHOSPHATE PRECIPITATION IN WASTEWATER TREATMENT. Environmental Protection Technology Series, EPA-R2-72-064. Washington, D.C.: EPA, December 1972. 95 p. Paperbound. Sold by Government Printing Office.

341g Morgan, W.E., and Fruh, E.G. AN INVESTIGATION OF PHOSPHO-ROUS REMOVAL MECHANISMS IN ACTIVATED SLUDGE SYSTEMS. Environmental Protection Technology Series, EPA-R2-72-031. Washington, D.C.: EPA, November 1972. 146 p. Paperbound. Sold by Government Printing Office.

341h Nesbitt, J.B. "Phosphorous Removal: The State of the Art." JOUR. WPCF 41(May 1969): 701-13.

341i Recht, H.L., and Ghassemi, M. PHOSPHATE REMOVAL FROM WASTE-WATERS USING LANTHANUM PRECIPITATION. FWQA Water Pollution Control Research Series, 17010 EFX. Washington, D.C.: FWQA, April 1970. 45 p. Paperbound. Sold by Government Printing Office.

Suspended Solids Removal

342 Burns and Roe, Inc. PROCESS DESIGN MANUAL FOR SUSPENDED SOLIDS REMOVAL. EPA Program no. 17030 GNO. Washington, D.C.: EPA, October 1971. 184 p.

A well-organized guide to the practical application of suspended solids removal processes. Gravity systems, including plain and chemical sedimentation, flotation, and tube settlers, are considered along with several filtration processes such as microstraining and deep-bed filtration. Operation and maintenance requirements for the various processes are delineated. Cost curves are provided to aid in making relative cost comparisons. Cost indexes are provided, but for some reason not all curves are indexed.

343   Culp, G.L., and Conley, W. "1, High Rate Sedimentation with the Tube Clarifier Concept" and "2, High Rate Filtration with the Mixed Media Concept." In WATER QUALITY IMPROVEMENT BY PHYSICAL AND CHEMICAL PROCESSES, pp. 149-66. Edited by E.F. Gloyna and W.W. Eckenfelder, Jr. Austin: University of Texas Press, 1970.

A complete yet somewhat biased review of the application of tube settlers to both water and wastewater treatment is presented in this two-part paper. Installation of tube settlers in new and existing facilities is discussed as well as process performance. In part 2, the general theory of in-depth filtration is presented in the section on mixed media filtration. Practical aspects are discussed through several examples of full-scale facilities. (For general review, see Appendix A.)

344   Tchobanoglous, G. "Filtration Techniques in Tertiary Treatment.: JOUR. WPCF 42(April 1970): 604-23.

The paper contains the results of experimental studies on the gravity filtration of secondary effluent. The performance of various filter bed designs was evaluated with and without chemical addition. The results presented are noteworthy because they provide insight into the pattern of solids removal within a filter for single, dual, and multi-media beds. The effect of polyelectrolyte addition on filtration is demonstrated.

a32   Weber, W.J., Jr. PHYSICOCHEMICAL PROCESSES FOR WATER QUALITY CONTROL. New York: John Wiley & Sons, 1972. Pp. 139-98.

An extensive chapter on filtration in which both the theoretical and practical aspects of filtration are delineated. (For general review, see Appendix A.) (Also cited in Section 3, B, E, F below, and H and Section 5, A.)

Suspended Solids Removal—Additional References

344a   Diaper, E.W.J. "Tertiary Treatment by Microstraining - Case Histories." WATER & SEWAGE WORKS 120(August 1973): 42-47.

344b   Libby, D.V.; Bell, G.R.; and Wirsig, O.A. "Chemical-Physical Treatment of Municipal Wastewater Using Continuous Countercurrent Filtration." JOUR. WPCF 44(April 1972): 574-82.

344c Lynam, B.T., and Bacon, V.W. "Filtration and Microstraining of Sec-
      ondary Effluent." In WATER QUALITY IMPROVEMENT BY PHYSICAL
      AND CHEMICAL PROCESSES, pp. 132-48. Edited by E.F. Gloyna and
      W.W. Eckenfelder, Jr. Austin: University of Texas Press, 1970.

344d Mixon, F.O. "Moving Bed Filtration of Municipal Waste." JOUR.
      WPCF 45(August 1973): 1718-27.

344e Yao, K.M., et al. Water and Wastewater Filtration: Concepts and Applica-
      tion." ENVIRONMENTAL SCIENCE & TECHNOLOGY 5(May 1971): 1105.

Dissolved Organics Removal

345 Garland, C.F., and Beebe, R.L. ADVANCED WASTEWATER TREAT-
     MENT USING POWDERED ACTIVATED CARBON IN RECIRCULATING
     SLURRY CONTACTOR CLARIFIERS. FWQA Water Pollution Control Re-
     search Series, ORD-17020 FKB. Washington, D.C.: EPA, July 1970.
     81 p. Paperbound. Sold by Government Printing Office.

The results of pilot studies conducted on powdered activated
carbon adsorption of secondary effluent are reported. This work
is significant because the studies represent some of the most extensive work to date on tertiary treatment using powdered activated carbon. It was found that polyelectrolyte flocculation was
required to settle the powdered carbon effectively. Two-stage
treatment achieved 60 to 84 percent chemical oxygen demand
(COD) removal with varying carbon doses. Costs are developed
for 10 mgd and 100 mgd systems.

346 Hamoda, M.F.; Brodersen, K.T.; and Sourirajan, S. "Organics Removal by
     low-Pressure Reverse Osmosis." JOUR. WPCF 45(October 1973): 2146-54.

A summary of the laboratory evaluation of newly developed osmotic membranes for low-pressure reverse osmosis is presented.
This work is noteworthy in that the practicality of using the
higher flux, lower pressure membranes to remove organic as well
as inorganic wastewater constituents was demonstrated. Treatment of primary effluent with reverse osmosis resulted in 96 percent COD removal, 72 percent nitrogen removal, and 100 percent phosphorus removal. Development of this low pressure
process could substantially reduce the cost of reverse osmosis.

347 Swindell-Dressler Co. PROCESS DESIGN MANUAL FOR CARBON AD-
     SORPTION. EPA Program no. 17020 GNR. Washington, D.C.: EPA,
     October 1971. 83 p.

This manual is devoted exclusively to granular activated carbon.
Coverage tends to be somewhat biased particularly in its empha-

sis on the advantages of complete physical-chemical treatment. Useful design considerations are presented in the areas of process and equipment design. A section on testing and selection of specific carbons for wastewater treatment is also included.

a32 Weber, W.J., Jr. PHYSICOCHEMICAL PROCESSES FOR WATER QUALITY CONTROL. New York: John Wiley & Sons, 1972. Pp. 199-259 and 330-35.

A comprehensive review of adsorption theory and factors influencing adsorption. A section on the theory and application of ultrafiltration for removal of high molecular weight organic materials is also included. (For general review, see Appendix A.) (Also cited in Section 3, B, E, F above, F below, and H and Section 5, A.)

Dissolved Solids Removal

348 Cruver, J.E., and Nusbaum, I. "Application of Reverse Osmosis to Wastewater Treatment." JOUR. WPCF 46(February 1974): 301-11.

Results obtained from extensive pilot plant studies conducted at the Pomona Water Renovation Plant in California are summarized in this paper. Although it is a case study, the operating experience gained through this project is the most extensive reported. Spiral-wound reverse osmosis systems were found to provide excellent rejection of inorganic salts and organics found in wastewater. Membrane fouling was identified as the major operational problem. Periodic chemical cleaning was necessary to maintain satisfactory product water flux. Several different methods of pretreatment for reverse osmosis were studied. On the basis of a useful cost comparison, reverse osmosis is shown to be competitive with other treatment methods.

349 Kreusch, E., and Schmidt, K. WASTEWATER DEMINERALIZATION BY ION EXCHANGE. EPA Water Pollution Control Research Series, 17040 EEE. Washington, D.C.: EPA, December 1971. 129 p. Paperbound. Sold by Government Printing Office.

The results of pilot studies conducted on domestic wastewater to determine the feasibility of using ion exchange procedures to demineralize wastewater are presented in this report. Prior to ion exchange treatment the wastewater was treated using lime clarification, dual media filtration, and carbon adsorption. The procedure found to be most reliable and recommended for application to wastewater demineralization is a strong acid cation exchange resin followed by a weak base anion exchange resin. A weak acid cation exchange resin preceeding the strong acid resin may be used to reduce operating costs. The operation and performance of several alternative ion exchange procedures are described, providing a rather complete review of basic procedures.

350 Merten, U., ed. DESALINATION BY REVERSE OSMOSIS. Cambridge, Mass.: The MIT Press, 1966. 289 p.

The eight chapters in this text are contributed by several prominent authors. Although the material presented is not focused specifically on wastewater treatment applications, it is the most complete reference on the reverse osmosis process. The principles and theory of osmosis and osmotic membranes are presented. Chapters on the preparation and evaluation of desalination membranes are also included. Engineering and operation of pilot and full-scale plants are discussed in separate chapters. The chapter on current activities is somewhat outdated now.

a32 Weber, W.J., Jr. PHYSICOCHEMICAL PROCESSES FOR WATER QUALITY CONTROL. New York: John Wiley & Sons, 1972. Pp. 261-362.

A thorough review of the theory and brief review of the application of the ion exchange process and membrane processes, including reverse osmosis, electrodialysis, and ultrafiltration. (For general review, see Appendix A.) (Also cited in Section 3, B, E, F above, and H and Section 5, A.)

Dissolved Solids Removal—Additional References

350a Environmental Systems Division Aerojet-General Corp. REVERSE OSMOSIS RENOVATION OF MUNICIPAL WASTEWATER. FWQA Water Pollution Control Research Series, 17040 EFQ. Washington, D.C.: FWQA, December 1969. 161 p. Paperbound. Sold by Government Printing Office.

350b Nusbaum, I.; Sleigh, J.H.; and Kremen, S.S. STUDY AND EXPERIMENTS IN WASTEWATER RECLAMATION BY REVERSE OSMOSIS. FWQA Water Pollution Control Research Series, 17040. Washington, D.C.: FWQA, May 1970. 116 p. Paperbound. Sold by Government Printing Office.

350c Smith, D.J., Eisenman, J.L. ELECTRODIALYSIS IN ADVANCED WASTE TREATMENT. FWPCA Publication, WP-20-AWTR-18. Washington, D.C.: FWPCA, February 1967. 219 p. Paperbound.

G. LAND TREATMENT PROCESSES

Land treatment involves the application of wastewater or effluents to the land for treatment on or in the soil matrix.

Wastewater Treatment

Documentation

* Importance of site characteristics
* Engineering hardware and analysis
* Suitable crops and crop yields
* Physics of water movement

Unresolved Issues

* Uptake of constituents by plants
* Extent of aerosol contamination
* Movement and survival of viruses
* Necessity for disinfection
* Degree of treatment required before application
* Trace element effects
* Operative removal mechanisms in the soil

General

351 McGauhey, P.H., and Krone, R.B. SOIL MANTLE AS A WASTEWATER TREATMENT SYSTEM. University of California SERL Report no. 67-11. Berkeley: University of California, December 1967. 201 p. Paperbound.

From an extensive literature review, the parameters by which the engineer and soil scientist might design a land application system are identified. The theory of wastewater treatment in soil is well covered and equations, such as for filtration, are developed. The authors conclude that there is enough information in the literature to permit the fully informed and imaginative engineer to exploit the soil mantle of the earth for wastewater treatment through engineered systems. However, criteria are sufficiently inexact that designs should provide land area for expansion of the system if necessary. Needed research areas are suggested, particularly on the infiltration and percolation tests.

352 Pound, C.E., and Crites, R.W. WASTEWATER TREATMENT AND REUSE BY LAND APPLICATION. 2 vols. Environmental Protection Technology Series, EPA - 660/2-73-006a. Washington, D.C.: EPA, August 1973. Vol. 1, 80 p.; vol. 2, 249 p. Paperbound. Sold by Government Printing Office.

In the summary report (vol. 1) the results of a nationwide study conducted on the current knowledge and techniques of land application are delineated. Factors involved in system design and operation are discussed for irrigation, overland flow, and infiltration-percolation approaches. In addition, evaluations were made of environmental effects, public health considerations and costs.

In the project report (vol. 2) literature is reviewed in detail and documentation is presented on selected sites visited. The relationship between climate and land application is probed. A history of several instances of abandonment of irrigation is included.

353 Reed, S.C. WASTEWATER MANAGEMENT BY DISPOSAL ON THE LAND. U.S. Army Corps of Engineers, Cold Regions Research and Engineering Laboratory, Special Report 171. Hanover, N.H.: U.S. Army Corps of Engineers, May 1972. 183 p. Paperbound.

Three land disposal techniques, spray irrigation, overland runoff, and rapid infiltration, are considered in this state-of-the-art review. Each technique is considered in detail including such aspects as wastewater characteristics, water quality goals, site conditions, operational criteria, and ecosystem response. The concept of renovative capacity is introduced in which the assumption is that there is a finite depth of soil in which major renovation occurs. The report is prepared by a multi-disciplined team including hydrologists, geologists, climatologists, soil scientists, and sanitary engineers. No cost information is included, but design components are discussed.

354 Stevens, R.M. GREEN LAND - CLEAN STREAMS: THE BENEFICIAL USE OF WASTEWATER THROUGH LAND TREATMENT. Philadelphia: Temple University, Center for the Study of Federalism, 1972. 330 p. Paperbound.

A proponent's view of land treatment of wastewater and sludge is presented. Numerous comparisons are raised between land treatment and advanced waste treatment. More than forty different facilities where land treatment or disposal is being practiced are described. The basis of these descriptions is primarily from the literature as well as from some questionnaires and interviews. The systems described are categorized as infiltration-type or overland flow-type. Data presented include costs, pollutant removal rates, and areas of land versus volumes of water applied. Social and political aspects of land treatment are discussed.

Irrigation

355 Muskegon County Board and Department of Public Works. ENGINEERING FEASIBILITY DEMONSTRATION STUDY FOR MUSKEGON COUNTY, MICHIGAN WASTEWATER TREATMENT - IRRIGATION SYSTEM. FWQA Water Pollution Control Research Series, 11010 FMY. Washington, D.C.: FWQA, September 1970. 174 p. Paperbound. Sold by Government Printing Office.

The Muskegon County project is one of the most controversial

Wastewater Treatment

land treatment systems in the country. Combined industrial and municipal wastewater is to be treated in aerated lagoons and spray irrigated on 10,000 acres in Michigan. Technical considerations of wastewater characteristics (with details on heavy metals), lagoon treatment, soils and groundwater, agriculture, and a simulation model for water quality and quantity in the lagoons are discussed in the report. Not included are cost evaluations, family displacement considerations, or information on the EPA and U.S. Army research efforts.

356   Parizek, R.R., et al. WASTEWATER RENOVATION AND CONSERVATION. The Pennsylvania State University Studies no. 23. University Park: Pennsylvania State University Press, 1967. 71 p. Paperbound.

Application of secondary effluent to croplands and forested areas at Penn State was initiated in 1962. This is the most comprehensive and well monitored research-field effort involving the spray irrigation of municipal wastewater. Approximately 80 percent of the effluent applied percolates through the soil and recharges the groundwater. Effluent was renovated when applied at rates of one, two, and four inches per week from April to December. Effluent was applied to some forested areas from November to April despite snow and ice buildup. The growth of trees, corn, and Reed canary grass has been measured and the soil water and groundwater monitored for changes in chemical quality.

357   Soil Conservation Service. "Soil-Plant-Water Relationship." In IRRIGATION. U.S. Department of Agriculture, SCS National Engineering Handbook, section 15. Washington, D.C.: U.S. Department of Agriculture, March 1964. 72 p.

This chapter, published separately, contains an excellent description of the physical properties of soil, soil water, and water intake into soil. Soil water consists of hyroscopic water that is unavailable to plants, capillary water that is available for plant growth, and gravitational or free water of which a certain amount is available for plant growth. Irrigation requirements commensurate with minimum plant needs may be determined using a series of graphs and charts. Rooting characteristics and irrigation techniques are discussed.

Overland Flow

358   Gilde, L.C., et al. "A Spray Irrigation System for Treatment of Cannery Wastes." JOUR. WPCF 43(August 1971): 2011-25.

The results obtained from a twelve-month research program conducted by Campbell Soup Co. at Paris, Texas, are presented. On the basis of this study, criteria are set forth for the various factors involved in overland flow or spray-runoff treatment.

52

The history of the spray-runoff approach to cannery wastewater treatment at Campbell Soup is traced from Napoleon, Ohio, to Chestertown, Maryland, and to Paris, Texas. Construction and operating costs are given and some of the operating problems are described.

359 Thomas, R.E. "Spray-Runoff to Treat Raw Domestic Wastewater." Presented at the International Conference on Land for Waste Management, October 1973, Ottawa, Canada. 15 p.

Field studies were conducted by the EPA at Ada, Oklahoma, to evaluate the capability of spray-runoff (overland flow) for year-round treatment of comminuted raw wastewater. The eighteen-month study period lasted from early 1971 through the summer of 1972. The comminuted raw wastewater was applied through nozzles attached to a rotatable boom about three feet above ground. Three different plots were used to test varying loading rates. The runoff from the three experimental plots was substantially better in quality than the effluent produced by conventional wastewater treatment processes. Nitrogen and phosphorus removal was better in summer than in winter; otherwise, there was little seasonal effect on performance.

Infiltration—Percolation

360 Bouwer, H.; Rice, R.C.; and Escarcega, E.D. RENOVATED SECONDARY SEWAGE BY GROUNDWATER RECHARGE WITH INFILTRATION BEDS. EPA Water Pollution Control Research Series, 16060 DRV. Washington, D.C.: EPA, March 1972. 102 p. Paperbound. Sold by Government Printing Office.

The feasibility of renovating secondary sewage effluent by groundwater recharge with infiltration basins was demonstrated at Flushing Meadows, Arizona. The five-year field project involved experimentation with loading rates, depth of inundation, various types of grassed basins, and periods of resting. The sandy two-acre test plot is located in the Salt River bed near Phoenix. Special efforts were made to promote nitrification and denitrification in the soil mantle by varying the periods of inundation and resting. Extensive monitoring was conducted and complete removals of BOD and fecal coliform were found.

361 McMichael, F.C., and McKee, J.E. WASTEWATER RECLAMATION AT WHITTIER NARROWS. California State Water Quality Control Board Publication no. 33. Sacramento: California State Water Quality Control Board, 1966. 100 p. Paperbound.

A very comprehensive report in which the infiltration-percolation system at Whittier Narrows, California, is documented. In an area (Rio Hondo) used for groundwater recharge with Colo-

rado River water, a test basin was built to study recharge with
activated sludge effluent. A second basin was built near the
treatment plant at Whittier Narrows. Extensive monitoring data
from both test basins are reported. With a surface layer of pea
gravel, loading rates up to ten feet per week have been sustained.

Land Treatment—Additional References

361a Environmental Protection Agency. PROCEEDINGS OF THE JOINT CON-
FERENCE ON RECYCLING MUNICIPAL SLUDGES AND EFFLUENTS ON
LAND. Washington, D.C.: National Association of State Universities
and Land-Grant Colleges, July 1973. 244 p. Paperbound.

361b Law, J.P.; Thomas, R.E.; and Myers, L.H. "Cannery Wastewater Treatment by High-Rate Spray on Grassland.: JOUR. WPCF 42 (September
1970): 1621-31.

361c Pair, C.H., ed. SPRINKLER IRRIGATION, 3rd ed. and Supplement.
Silver Spring, Md.: Sprinkler Irrigation Association, 1969 and 1973.
444 p. and 101 p.

361d U.S. Department of Agriculture, Soil Conservation Service. DRAINAGE
OF AGRICULTURAL LAND. Port Washington, N.Y.: Water Information
Center, 1973. 430 p. Paperbound.

361e Zimmerman, J.P. IRRIGATION. New York: John Wiley & Sons,
1966. 516 p.

H. DISINFECTION

The process of disinfection has been a subject of renewed interest in the field
of wastewater treatment. In spite of the widespread and long-used practice of
wastewater disinfection, almost exclusively by chlorination, knowledge of the
process as a whole is still in its infancy. Recently gained knowledge of the
chemistry and the process kinetics of wastewater chlorination has led to a re-
alization that many current chlorination operations are too inadequately designed
and operated to achieve consistent disinfection efficiency. A fundamental
difference of opinion still remains regarding the modes of germicidal activity
and the specific chemical forms involved during the disinfection process. Recognition of the need to control viruses in wastewater discharges and domestic
water supplies has led to an emphasis in research on virus inactivation and
removal. Achieving virus control is likely to have a profound effect on the
methods of disinfection employed and on the means of controlling disinfection
efficiency.

Documentation

* History of chlorination
* Qualitative description
* Generalized design criteria
* Application and operation of chlorination
* Chemistry of chlorination
* Factors controlling disinfection efficiency

Unresolved Issues

* Effect of constituents on chlorination efficiency
* Optimum design of chlorination facilities
* Adverse effects of by-products from chlorination
* Modes of germicidal action on bacteria and virus
* Application of disinfection methods other than chlorination
* Treatment requirements for virus inactivation
* Monitoring for virus control

General

362 Chang, S.L. "Modern Concepts of Disinfection." In PROCEEDINGS OF THE NATIONAL SPECIALTY CONFERENCE ON DISINFECTION, pp. 635-81. New York: ASCE, 1970.

> This paper provides an excellent guide to the scientific under-standing of the disinfection process. Such an understanding re-quires knowledge in four scientific disciplines: (1) physicochem-istry of the disinfectant, (2) cyto-chemical nature and physical state of the pathogen, (3) reaction of the two as a rate process, and (4) quantitative effects of factors in the reaction medium. All groups of disinfecting agents and pathogens are reviewed on the above basis. Methods of treating survival rate data to ac-count for deviations from first-order reactions are presented. The collected papers in which this study appears provide an excellent review of the current state of knowledge of disinfec-tion.

a11 Fair, G.M.; Geyer, J.C.; and Okun, D.A. WATER AND WASTEWA-TER ENGINEERING. Vol. 2. New York: John Wiley & Sons, 1968. Pp. 31-1 to 31-29.

> The classical theory of chemical disinfection is presented, along with a brief review of alternative methods of disinfection. The review of chlorination chemistry is especially concise. (For general review, see Appendix A.) (Also cited in Section 3, A, B, and E.)

363 Morris, J.C. "Chlorination and Disinfection: State of the Art."

JOUR. AWWA 63(December 1971): 769-74.

The author points out that disinfection practice has not taken advantage of new scientific knowledge because of a lack of incentive for positive change. The need to achieve virus inactivation through disinfection should stimulate adaptation of new technology to full-scale operations. The effect of pH on chlorine derivatives during breakpoint chlorination is discussed. The major engineering considerations required for efficient chlorination are discussed. A complete and objective review of the application of other disinfectants is presented as well as a summary of research needs.

a32   Weber, W.J., Jr. PHYSICOCHEMICAL PROCESSES FOR WATER QUALITY CONTROL. New York: John Wiley & Sons, 1972. Pp. 413-56.

A brief review is given on all disinfection methods, although emphasis is placed on chlorine chemistry. (For general review, see Appendix A.) (Also cited in Section 3, B, E, and F and Section 5, A.)

364   White, G.C. HANDBOOK OF CHLORINATION. New York: Van Nostrand Reinhold Co., 1972. 744 p.

The book offers one of the most comprehensive discussions available on the application and operation of the chlorination process. There are thirteen chapters that cover the history and manufacturing of chlorine, the design of chlorination facilities, and the operation and maintenance of chlorination equipment as well as the application of chlorine to water and wastewater. The chapter on the chemistry of chlorination is thorough and well referenced, particularly with regard to the reactions of chlorine with nitrogenous compounds. Separate chapters are devoted to chlorine dioxide and hypochlorination. There is a concise evaluation of several alternative methods of disinfection: ozone, bromine, iodine, silver, and ultra violet rays; however, the evaluation is somewhat biased toward the use of chlorine.

Chlorination

365   Collins, H.F.; Selleck, R.E.; and White, G.C. "Problems in Obtaining Adequate Sewage Disinfection." In PROCEEDINGS OF THE NATIONAL SPECIALTY CONFERENCE ON DISINFECTION, pp. 137-61. New York: ASCE, 1970.

Findings derived from laboratory research are presented. Based on these findings, the chlorination system design factors necessary to achieve efficient disinfection are identified. In particular, these include provision for rapid initial mixing of chlorine solution and wastewater and the use of plug flow type contact-chambers. Additional experimental results are presented from which it is concluded that free chlorine residual is not respon-

sible for disinfection during the initial mixing process. This is
significant in that it disagrees with findings of others (See 363,
369.) The bactericidal activity of combined residuals versus
time is also discussed.

366   Joint Committee of the Water Pollution Control Federation and American
      Society of Civil Engineers. SEWAGE TREATMENT PLANT DESIGN.
      Manual of Practice no. 8. Washington, D.C.: 1972. 375 p. Paper-
      bound.

      A practical guide to the application of chlorination to sewage
      treatment. Average dosage requirements for disinfection and
      other chlorine use are discussed. Virus inactivation is not con-
      sidered. A brief review of the important considerations in de-
      signing chlorine equipment installations and in operating the
      equipment is presented. This manual is intended as reference for
      design engineers and treatment plant operators. (Also cited in
      Section 3, C.)

a21   Metcalf & Eddy, Inc. WASTEWATER ENGINEERING: COLLECTION,
      TREATMENT, DISPOSAL. San Francisco: McGraw-Hill, 1972. Pp.
      353-63 and 470-78.

      Disinfection theory and chlorine chemistry are reviewed briefly
      as part of one chapter. A useful review of chlorination appli-
      cations and design considerations for chlorination facilities is
      presented in a separate chapter. (For general review, see Ap-
      pendix A.) (Also cited in Section 2, A; Section 3, A, B, C,
      E, and F; Section 4, B; and Section 5, A.)

367   Morris, J.C. "Future of Chlorination." JOUR. AWWA 58(November
      1966): 1475-81.

      The development of chlorination practice, traced to the present
      time, is followed by a brief but incisive analysis of the direc-
      tion that future research should take. A useful table is pre-
      sented in which the concentrations of the various chlorine de-
      rivatives required to provide inactivation of enteric bacteria,
      viruses, cysts, and bacterial spores are reported. Significant
      conclusions are drawn about future chlorination practice, the
      most significant being that virus control rather than bacteria
      control will be the governing factor in disinfection. Possible
      indexes to replace or supplement the coliform index are dis-
      cussed.

a29   Sawyer, C.N., and McCarty, P.L. CHEMISTRY FOR SANITARY ENGI-
      NEERS. 2nd ed. San Francisco: McGraw-Hill, 1967. Pp. 363-75.

      A very brief and simplified review of chlorine chemistry is pro-
      vided, followed by a description of several methods of deter-
      mining chlorine residuals. Good from an introductory stand-
      point. (For general review, see Appendix A.) (Also cited

in Section 3, B and J.)

Virus Removal

368   Berg, G. "An Integrated Approach to the Problem of Virus in Water."
In PROCEEDINGS OF THE NATIONAL SPECIALTY CONFERENCE ON
DISINFECTION, pp. 339-64. New York: ASCE, 1970.

     A discussion of the epidemiology of viruses in water is present-
ed. The need to provide total virus removal from any water
man might consume is emphasized. Several potential methods
of quantitatively recovering viruses from large volumes of water
are described and the difficulties of achieving such virus recovery are noted. A useful summary of viral removal efficiency
provided by conventional and advanced wastewater treatment
methods is provided with discussion. A brief review of the
available methods for virus disinfection is given. The in-depth
discussion on the direction of future research provides an excellent perspective on the subject of virus removal.

369   Kruse, C.W.; Olivieri, V.P.; and Kawata, K. "The Enhancement of
Viral Inactivation by Halogens." WATER AND SEWAGE WORKS
118(June 1971): 187-93.

     A discussion is presented on the need to upgrade current disin-
fection practices to assure removal of viruses from sewage effluents, despite lack of epidemiological evidence to support such
need. A brief review of the chemistry of halogens in water is
given, followed by a discussion of the possible modes of germicidal action of the various halogens. Interesting and significant
findings of experiments are reported from which it is concluded
that significant viral inactivation can be achieved in sewage
effluents without breakpoint chlorination, provided the halogen
is "flash-mix" with the sewage. This finding is consistent with
the hypothesis that viral inactivation is largely by free halogens
that exist momentarily during "flash-mixing." The dramatic effect of $_pH$ on viral inactivation efficiency is also documented.

370   Pavon, J.L., et al. "Virus Removal From Wastewater Using Ozone."
WATER & SEWAGE WORKS 119(December 1972): 59-67.

     An informative and concise background and literature review is
presented on virus content and viability in wastewater and the
practical methods for determination of virus in wastewater. Virus removals by conventional primary and secondary treatment
processes are reviewed. A literature review of the efficiency
of disinfection of viruses by halogens is presented with the conclusion that current chlorination standards for conventional
wastewater treatment plants do not assure a virus-free effluent.
An experiment to determine the effectiveness of ozone as a
viral disinfectant of secondary effluent is described. For the

test virus used, the rate of inactivation with ozone was greater for virus than for bacteria. Although the article is directed towards building a case for ozonation, useful background information is provided.

Other Methods

371 Eliassen, R., and Trump, J. "High-Energy Electrons Offer Alternative to Chlorine." CALIFORNIA WATER POLLUTION CONTROL ASSOCIATION BULLETIN 10(January 1974): 50-54.

A practical method of using accelerated electron irradiation for wastewater and sludge disinfection is described. The paper is interesting and is on a general and nontechnical level. A fairly complete listing of technical references is provided. A crude economic analysis is included in which this system is found to be competitive with chlorination.

372 Rosen, H.M. "Use of Ozone and Oxygen in Advanced Wastewater Treatment." JOUR. WPCF 45(December 1973): 2521-36.

A concise but somewhat biased review of ozone applications. A useful review of the process of ozone generation, including descriptions of commercially available ozonators, is presented. Several wastewater ozonation systems are discussed, based on this background. This paper contains a good compilation of background information on the subject of ozone application for disinfection as well as for other advanced treatment processes.

Disinfection—Additional References

372a Berg, G. TRANSMISSION OF VIRUS BY THE WATER ROUTE. New York: Interscience Publishers, 1966. 484 p.

372b Chaudhuri, M., and Engelbrecht, R.S. "Virus Removal in Wastewater Renovation by Chemical Coagulation and Flocculation." In ADVANCES IN WATER POLLUTION RESEARCH: PROCEEDINGS OF THE 5TH INTERNATIONAL CONFERENCE, pp. II-20/1 to II-20/21. Edited by S.H. Jenkins. New York: Pergamon Press, 1971.

372c Majumdar, S.B.; Ceckler, W.H.; and Sproul, O.J. "Inactivation of Poliovirus in Water by Ozonation." JOUR. WPCF 45(December 1973): 2433-43.

372d Nebel, C., et al. "Ozone Disinfection of Industrial-Municipal Secondary Effluents." JOUR. WPCF 45(December 1973): 2493-2507.

372e State of California Department of Public Health, Bureau of Sanitary Engineering. WASTEWATER CHLORINATION FOR PUBLIC HEALTH PRO-

TECTION. PROCEEDINGS OF THE FIFTH ANNUAL SANITARY ENGI-
NEERING SYMPOSIUM. Berkeley: May 1970. 131 p. Paperbound.

372f Wolf, H.W. "Bacterial & Viral Control & Water Reuse." In AD-
VANCED WASTE TREATMENT AND WATER REUSE SYMPOSIUM, DAL-
LAS, TEXAS. Washington, D.C.: EPA, January 1971.

I. TREATMENT PLANT OPERATION AND MAINTENANCE

Treatment plant operation involves the labor requirements, duties, and organization of treatment plant personnel. Maintenance involves routine inspection
and repair of equipment and structures, plus record keeping of maintenance
activities.

Documentation

* Effects of constituents on equipment
* Equipment maintenance requirements
* Manpower requirements

Unresolved Issues

* Optimum methods of conventional process control
* Relationship of number of operators to plant performance
* Methods of operation for advanced treatment processes
* Application of computer control

373 Garber, W.F. "Treatment Plant Equipment and Facilities Maintenance."
JOUR. WPCF 42(October 1970): 1740-70.

Maintenance requirements in sewage treatment plants are described through reference to several large city systems. Manpower needs and task descriptions are reviewed. Several examples of treatment plant staff organizational structures are presented. Design considerations for maintenance facilities are discussed with reference to existing facilities. Some details of in-
ternal control of maintenance and repair activities are described.
Examples of work order forms and maintenance record forms are
included. An enlightening paper applicable to both large and
small plants.

374 Environmental Protection Agency. ESTIMATING STAFFING REQUIRE-
MENTS FOR MUNICIPAL WASTEWATER TREATMENT FACILITIES. EPA
Contract no. 68-01-0328. Washington, D.C.: March 1973. 81 p.
Paperbound. Sold by Government Printing Office.

Staff requirements for both conventional and advanced wastewater

treatment processes ranging from 0.5 to 25 mgd can be estimated
using this manual. Staff requirement curves are provided for
each type of process as well as supervisory and support func-
tions. Task descriptions are provided for the various treatment
plant personnel. This reference is particularly valuable in the
planning stages of a treatment plant.

375   Water Pollution Control Federation, Subcommittee on Operation of Wastewater Treatment Plants. OPERATION OF WASTEWATER TREATMENT
PLANTS. Manual of Practice no. 11. Washington, D.C.: 1970. 195
p.

This manual is designed as both a teaching and working reference for the treatment plant operators. Coverage is limited to
conventional, primary, and secondary treatment systems. Advanced wastewater treatment is not considered. A brief review
of the fundamentals and a description of the conventional processes are presented. Common operating problems with each
process are identified, and procedures are suggested to solve the
various problems. Sampling procedures are described.

J.   WASTEWATER ANALYSIS AND MONITORING

Monitoring and analysis involve the sampling of waste streams and the methods
of laboratory analysis to. determine wastewater constituents. Monitoring data
and information are necessary for treatment process control as well as for assessment of water quality for planning or enforcement of pollution control requirements.

Documentation

*   Analytical methods
*   Sampling procedures and techniques
*   Storage of samples

Unresolved Issues

*   Sampling frequency
*   Long-term reliability and stability of monitoring equipment
*   Constituents to be monitored
*   Use of monitoring data in process control
*   Sampling and analysis for virus identification
*   Indicator organisms for virus control monitoring

376   Associated Water and Air Resource Engineers, Inc. HANDBOOK FOR
MONITORING INDUSTRIAL WASTEWATER. Washington, D.C.: EPA,
1973. 152 p.

Although prepared for use in monitoring industrial wastewaters, this manual is a useful reference for any type of water or wastewater monitoring.

377 Mancy, K.H., ed. INSTRUMENTAL ANALYSIS FOR WATER POLLU-
TION CONTROL. Ann Arbor, Mich.: Ann Arbor Science Publishers, Inc., 1971. 331 p.

The principles and applications of instrumental measurement of water and wastewater constituents are described. The book is intended to be a general reference for those with or without chemistry backgrounds. Automated analytical methods for specific constituents are discussed.

a29 Sawyer, C.N., and McCarty, P.L. CHEMISTRY FOR SANITARY EN-
GINEERS. 2nd ed. San Francisco: McGraw-Hill, 1967. Pp. 285-499.

A review of the sources and significance of the major components of wastewater is given, including an excellent discussion of analytical procedures. This is a good source of information on the basic level. (For general review, see Appendix A.) (Also cited in Section 3, B and H.)

378 Taras, M.J., et al, eds. STANDARD METHODS FOR THE EXAMINA-
TION OF WATER AND WASTEWATER. 13th ed. New York: American Public Health Association, American Water Works Association, Water Pollution Control Federation, 1971. 874 p.

A basic procedural manual for analysis of water and wastewater. Techniques for the determination of physical, chemical, and biological properties of water and wastewater are included in this comprehensive reference. Bioassay procedures are included, and methods for collecting and storing samples are presented.

Section 4

EFFLUENT REUSE AND DISPOSAL

Section 4

EFFLUENT REUSE AND DISPOSAL

After treatment, wastewater may be put to sequential use or returned to surface waters or groundwaters. In the past, reuse of effluent has been constrained, in many areas, by an abundance of fresh water, by lack of knowledge concerning health effects, and by lack of public acceptance. Various innovative reuse applications for treated effluent are stressed in this section. More conventional disposal methods to receiving rivers, lakes, estuaries, and oceans, and to the land are also documented.

A. EFFLUENT REUSE

Treated effluent can be used in many beneficial ways: in the production of agricultural and forest products, in the production of fish and aquatic life, to recharge groundwater aquifers, for the development of recreational lakes, and for industrial process and cooling water.

Documentation

* Qualitative description of potential reuse applications
* General information on health effects

Unresolved Issues

* Constituent quality required in alternative applications
* Need for reuse
* Health effects of specific constituents and viruses
* Public acceptance

General

401 Environmental Protection Agency. LAND APPLICATION OF SEWAGE EFFLUENTS AND SLUDGES: SELECTED ABSTRACTS. Environmental Protection Technology Series, EPA - 660/2-74-042. Corvallis, Oreg.:

June 1974. 249 p. Paperbound. Sold by Government Printing Office.

Included in this report are 568 selected abstracts that were compiled as a result of several recent EPA contracts. A thorough subject, location, and author index precedes the abstracts, which are grouped chronologically and identified as to whether emphasis is on effluents, sludges, or industrial wastes.

a12   Gloyna, E.F., and Eckenfelder, W.W., Jr., eds. WATER QUALITY IMPROVEMENT BY PHYSICAL AND CHEMICAL PROCESSES. Austin: University of Texas Press, 1970. Pp. 3-79.

The first part of this book, which comprises individual papers, contains a useful analysis of water quality requirements for reuse. All major reuse applications are documented, along with considerations of health aspects and objectives of reuse. (For general review, see Appendix A.) (Also cited in Section 5, A.)

Agricultural

402   Bendixen, T.W., et al. "Ridge and Furrow Liquid Waste Disposal in a Northern Latitude." JOUR. SAN. ENGR. DIV. ASCE 94(February 1968): 147-57.

The efficiency of land disposal under year-round operation, with emphasis on cold climate effects, was investigated in pilot scale tests and a full-scale operation. The pilot tests were used to compare sprinkling, ridge and furrow, and flood irrigation methods. These were conducted near the EPA's Cincinnati Research Center using septic tank effluent. Data are given on system longevity and removal efficiencies.

For a full-scale test, the six-year-old ridge and furrow disposal system serving the city of Westby, Wisconsin, was chosen. Site characteristics are described and system efficiency is detailed.

403   Law, J.P. NUTRIENT REMOVAL FROM ENRICHED WASTE EFFLUENT BY THE HYDROPONIC CULTURE OF COOL SEASON GRASSES. FWQA, Water Pollution Control Research Series, 16080. Washington, D.C.: FWQA, October 1969. 33 p. Paperbound. Sold by Government Printing Office.

Tall fescue and perennial rye grass were grown in hydroponic culture tanks to evaluate their nutrient removal capabilities when supplied with secondary sewage effluent. Hydroponic culture involves providing nutrients for plant growth solely from the liquid applied. Six tanks were filled with a 15-inch bed of pea gravel covered with a 2-inch layer of coarse sand. In control tanks, nitrogen removal was 56 percent and phosphorus removal was 5 percent. With grass growth, nitrogen removal was 78 percent and phosphorus removal was 13 percent. Yields and uptake calculations are also given.

404   Young, R.H.F., et al. "Wastewater Reclamation by Irrigation." JOUR.
      WPCF 44(September 1972): 1808-14.

      In this study, which is being planned for central Oahu, data
      pertinent to Hawaii regarding transmission of viruses and dissolved
      minerals through soils under controlled field conditions will
      be developed. Wastewater effluent, after undergoing various
      forms of pretreatment, will be applied to irrigate grass sod and
      sugar cane. The information gained from this study will be used
      to analyze the expected degree of health hazard and mineral
      build-up in the receiving groundwater under various operating
      conditions and with various degrees of pretreatment.

405   Sullivan, R.H., et al. SURVEY OF FACILITIES USING LAND APPLICA-
      TION OF WASTEWATER. EPA - 430/9-83-006. Washington, D.C.:
      EPA, July 1973. 365 p. Illustrations. Paperbound. Sold by Govern-
      ment Printing Office.

      The results of a 1972 field survey of one hundred domestic and
      industrial systems using land application of wastewater are pre-
      sented in this report. Most of the sites visited were using irri-
      gation, and the data are treated statistically using five climatic
      zones for the United States. Abstracts from foreign experience
      are included. Photographs of many of the systems are included,
      and a survey was also made of state health regulations. The
      appendix material is nearly as valuable as the body of the re-
      port, containing all the raw data, the trip narratives, results of
      a parallel mail survey, and two background papers.

Silvicultural

406   Kardos, L.T., and Sopper, W.E. "Renovation of Municipal Wastewater
      through Land Disposal by Spray Irrigation." In RECYCLING TREATED
      MUNICIPAL WASTEWATER AND SLUDGE THROUGH FOREST AND CROP-
      LAND, pp. 148-63. Edited by W.E. Sopper and L.T. Kardos. Univer-
      sity Park: The Pennsylvania State University Press, 1973.

      This is one of thirty-two papers given at a major symposium on
      land treatment in August 1972. Many of the papers deal with
      aspects of the ten-year-old "living filter" experiment involving
      forest and cropland at Penn State. Data are presented on the
      concentrations of nitrogen, phosphorus, boron, and other chem-
      icals added to the land. Concentrations of phosphorus and ni-
      trate at various depths in the soil are reported for individual
      areas. In most cases, for forested areas, the application rates
      were less than for corn or reed canary grass to produce the same
      concentration of nutrients in the soil water.

407   Sutherland, J.C., et al. "Irrigation of Trees and Crops with Sewage
      Stabilization Pond Effluent in Southern Michigan." In WASTEWATER USE
      IN THE PRODUCTION OF FOOD AND FIBER - PROCEEDINGS, EPA -

660/2-74-041, pp. 295-313. Washington, D.C.: EPA, June 1974. Paperbound. Sold by Government Printing Office.

The authors found that the use of treated wastewater for irrigation of hardwood cuttings and seedlings has produced considerable increases in survival and height growth over that of unirrigated stock. The growth of established red pine was not increased significantly, however, and nutrient analysis of foliage has led the authors to be concerned about levels of boron that may lead to toxicity conditions in future years. Related research on silviculture in Michigan is described.

Aquacultural

408 Coleman, M.S., et al. "Aquaculture as a Means to Achieve Effluent Standards." In WASTEWATER USE IN THE PRODUCTION OF FOOD AND FIBER - PROCEEDINGS, EPA - 660/2-74-041, pp. 199-214. Washington, D.C.: EPA, June 1974. Paperbound. Sold by Government Printing Office.

The Quail Creek lagoon system north of Oklahoma City, an experimental one mgd system, was the site of this research work on aquaculture. In an attempt to demonstrate the potential for upgrading the quality of lagoon effluent by biological means, shiners, minnows, and catfish were introduced into the last five of a series of six ponds. The author denotes the increased quality of the contents of the ponds from beginning to end and gives the yields of fish harvested over a six-month period. The preliminary economics of the system are encouraging.

409 Kerfoot, W.B., and Jacobs, S.A. "Permissible Levels of Heavy Metals in Secondary Effluent for Use in a Combined Sewage Treatment - Marine Aquaculture System." In WASTEWATER USE IN THE PRODUCTION OF FOOD AND FIBER - PROCEEDINGS, EPA - 660/2-74-041, pp. 79-101. Washington, D.C.: EPA, June 1974. Paperbound. Sold by Government Printing Office.

Pilot scale research on the use of sewage-grown algae for feeding of caged oysters is documented. Because oysters concentrate heavy metals, the algae-laden secondary effluent was diluted with sea water that was very low in metals. Copper, cadmium, nickel, chromium, lead, and zinc were studied, and suggested levels for satisfactory culture are given. This exciting research is continuing at Woods Hole Oceanographic Institution, Massachusetts.

410 Wong, B.R.; Colt, J.E.; and Tchobanoglous, G. WATER QUALITY MANAGEMENT IN FISH CULTURE, WITH SPECIFIC REFERENCE TO CHANNEL CATFISH: AN ANNOTATED BIBLIOGRAPHY. Davis: Department of Civil Engineering, University of California, June 1974.

68 p. Paperbound.

Selected references dealing with maintenance of water quality in managed aquaculture systems are presented. Although water quality is stressed, general aquaculture references and articles on biological and physical aspects of culture are included. The subject area is rounded out with texts and articles on economics and legislation pertinent to fish culture with emphasis on catfish farming. The reader is given a brief background on each subject included and is given the highlights of documentation in the literature on unresolved or controversial issues.

Groundwater Recharge

411 Bouwer, H. "Groundwater Recharge Design for Renovating Wastewater." JOUR. SAN. ENGR. DIV. ASCE 96(February 1970): 59-74.

The design concept of a groundwater recharge system at Flushing Meadows near Phoenix, Arizona, is described. The system consists of six parallel, rectangular recharge basins with wells located midway between the basins and on both sides. Activated sludge effluent is dosed intermittently to the basins.

Using an analog technique, the horizontal and vertical hydraulic conductivity of the underlying aquifer is determined from the recharge rate and the water level response in two observation wells. From the hydraulic conductivity data thus obtained, an effective transmissibility coefficient can be computed for a multiple-basin, multiple-well recharge and recovery system.

412 Wesner, G.M., and Baier, D.C. "Injection of Reclaimed Wastewater into Confined Aquifers." JOUR. AWWA 62(March 1970): 203-10.

Three multi-casing injection wells and thirteen multi-casing observation wells were constructed to determine the feasibility of injecting treated wastewater into a confined aquifer. This project was undertaken to halt salinity intrusion from the ocean into potable aquifers in Orange County, California. The injected effluent was of tertiary quality. Coliform bacteria have been found to travel one hundred feet, but no viruses were found in the injection water or wells. Many chemical constituents did not move conservatively in the aquifer. Hardness and oxygen-demanding materials were reduced significantly by travel in the confined aquifer. Some highly resistant, soluble organic materials have caused taste and odor problems in water samples withdrawn from observation wells 545 feet away.

Recreational

413 Martin, B. "Sewage Reclamation at Golden Gate Park." SEWAGE AND INDUSTRIAL WASTES 23(March 1951): 319-20.

The sewage treatment system at Golden Gate Park in San Francisco and the irrigation of the park grounds with plant effluent are described. Treatment consists of activated sludge plus chlorination. The chlorination is controlled to maintain a residual of two milligrams per liter. Effluent is blended with well water prior to irrigation of the park grounds and maintenance of the water levels of recreational lakes. Costs are given that show imported irrigation water is twice as expensive as reclaimed water.

414 Merrell, J.C., et al. THE SANTEE RECREATION PROJECT, SANTEE, CALIFORNIA, FINAL REPORT. FWPCA Water Pollution Control Research Series, WP-20-7. Cincinnati: FWPCA, 1967. 165 p. Paperbound.

The results of a study of the Santee, California, recreational lakes, which were deliberately planned to utilize the reclaimed wastewater of the community, are presented. This seven-agency cooperative study evaluates the fate of virus, total and fecal coliform, and fecal streptococci as wastewater passed through secondary treatment, infiltration-percolation, and the recreational lakes. Observations related to the eutrophication, vector control, epidemiology, and the social acceptance and ecology of the entire recreational park were obtained. No health hazards have been demonstrated in three years of virus studies; this has helped to promote public acceptance of the project.

415 Storey, E.H., and Ditton, R.B. "Water Quality Requirements for Recreation." In WATER QUALITY IMPROVEMENT BY PHYSICAL AND CHEMICAL PROCESSES, pp. 57-63. Edited by E.F. Gloyna and W.W. Eckenfelder, Jr. Austin: University of Texas Press, 1970.

Quality requirements for recreation such as physical, biological, chemical, and psychological variables are discussed, with emphasis on public health. The requirements for the first three variables are referenced to two key documents on water quality criteria. The authors consider aesthetics and the public's perception of water pollution to be of overriding importance.

Industrial

416 Horstkotte, G.A., et al. "Full-Scale Testing of Water Reclamation System." JOUR. WPCF 46(January 1974): 181-97.

Although this is a case study of a California installation, it represents a large, planned reclamation system for industrial reuse. The sanitary and water districts of Contra Costa County have signed a contract that will enable industries to use 75 percent of the 19 mgd (in 1975) of reclaimed water for cooling water and the remainder for process water. The testing of this tertiary treatment plant is reported in depth, and data on effluent quality are included. The contractual water quality limits as listed require both phosphorus and nitrogen reductions in the tertiary works.

417 Petrasek, A.C., Jr., and Esmond, S.E. "Industrial Uses of Municipal Wastewater." INDUSTRIAL WATER ENGINEERING 10(July/August 1973): 26-29.

Water quality requirements for industrial uses ranging from cooling water to needs for textile products are presented and discussed. Industrial uses are cited for various quality levels ranging from primary effluent to filtered, activated carbon effluent. All the cited levels of quality are referenced in a comprehensive table of quality parameters from the water reclamation plant at Dallas, Texas.

Obstacles

418 Benarde, M.A. "Land Disposal and Sewage Effluent: Appraisal of Health Effects of Pathogenic Organisms." JOUR. AWWA 65(June 1973): 432-40.

In this article, the author reconstructs the steps by which researchers have studied the effects of irrigation with undisinfected wastewater on bacterial survival and disease outbreaks. Extensive literature references are reviewed to evaluate the prevalence of diseases among workers handling wastewater as irrigation water. The literature on viral diseases as related to potable water is also reviewed, and the entire health effects issue is placed in excellent perspective. Various public health standards for disinfection prior to land application are given.

419 Bruvold, W.H., and Ward, P.C. "Using Reclaimed Wastewater: Public Opinion." JOUR. WPCF 44(September 1972): 1690-96.

Public opinion is often cited as an obstacle to wastewater reclamation and reuse. The authors have investigated public attitudes in California through extensive interviews. Using five communities with reuse applications and five communities of comparable size without, the authors studied public knowledge, beliefs, preferences, and reasons for opposing reuse. A key table is presented in which twenty-five uses of reclaimed water are listed in order of public opposition. The authors conclude that the least objectionable uses should be implemented first.

420 Drewry, W.A., and Eliassen, R. "Virus Movement in Groundwater." JOUR. WPCF 40(August 1968): pt. 2, R257-71.

Retention mechanisms for virus in soil were studied using radio-isotope-tagged viruses. The major mechanism was found to be adsorption. Adsorption increased with increasing clay content of soils and with increasing cation concentrations. Adsorption

is greatly affected by soil water pH, being greatest at pH 7.0
to 7.5 and decreasing at higher pH values. From the results of
this extensive experimental study, it was concluded that virus
movement through saturated soils should present little hazard to
underground water supplies, provided that the soil strata are
continuous.

421   Sepp, E. THE USE OF SEWAGE FOR IRRIGATION, A LITERATURE RE-
      VIEW. Berkeley: California State Department of Public Health, Bureau
      of Sanitary Engineering, 1971. 41 p. Paperbound.

      This is a comprehensive literature review on the public health
      aspects of irrigation with wastewater. Five major areas are
      considered: (1) disease outbreaks caused by sewage irrigation,
      (2) survival of pathogenic organisms and parasites on vegetables
      and in soil, (3) occurrence of pathogens and parasites in pol-
      luted water, (4) sewage treatment and distribution for irrigation,
      and (5) purifying effects of irrigation on the percolating water.
      A survey of public health standards for sewage irrigation used
      in the western states and in European countries was conducted.
      Studies of aerosol dispersion and related coliform presence in
      aerosols are reviewed. The bibliography consists of 138 refer-
      ences, some dating back to 1912.

B. EFFLUENT DISPOSAL IN RECEIVING WATERS

Minimization of toxic effects on aquatic life and public health and negative
aesthetic impact are important considerations in the disposal of effluents in
receiving waters. Various prediction techniques and mathematical models are
available which can be used to estimate the effects of dilution, dispersion,
and decay on effluent constituent concentrations in streams, lakes, estuaries,
and oceans.

Documentation

*   Outfall and diffuser design
*   Mathematical models for prediction of waste transport
*   Qualitative description of discharge systems
*   General effects of nitrogen and phosphorus
*   Effects of discharges on assimilative capacity
*   Prediction of effluent dispersion and diffusion

Unresolved Issues

*   Specific elements critical to eutrophication
*   Toxic effects of specific constituents
*   Prediction of constituent interaction in receiving water
*   Degree of treatment required to maintain water quality

General

a16   McGauhey, P.H. ENGINEERING MANAGEMENT OF WATER QUALI-
      TY. New York: McGraw-Hill, 1968. Pp. 233–62.

      Separate chapters are included on water quality management in
      estuaries, oceans, and fresh water. Dispersion of wastes, oxy-
      gen sag calculations, and eutrophication are discussed. (For
      general review, see Appendix A.) (Also cited in Section 4, C.)

a21   Metcalf & Eddy, Inc. WASTEWATER ENGINEERING: COLLECTION,
      TREATMENT, DISPOSAL. San Francisco: McGraw-Hill, 1972. Pp.
      673–705.

      Effluent disposal into lakes, rivers, estuaries, and oceans is dis-
      cussed and illustrated with examples. (For general review, see
      Appendix A.) (Also cited in Section 2, A; Section 3, A, B,
      C, E, F, and H; and Section 5, A.)

Stream Sanitation

422   Parker, F.L., and Krenkel, P.A., eds. ENGINEERING ASPECTS OF
      THERMAL POLLUTION. Nashville, Tenn.: Vanderbilt University Press,
      1969. 351 p.

      Alternative solutions for heated water discharge are contained
      in the proceedings of this second symposium on thermal pollution
      (the first symposium dealt with biological aspects). Because it
      was impossible to separate completely biological and engineering
      aspects, many of the discussions of the twelve papers presented
      considered the temperature tolerances of fish. Types of cooling
      devices, flowsheets for recycling heated water, and modeling of
      heated water discharges are discussed, but the most controversial
      subject was the establishment of water-quality standards for tem-
      perature.

a24   Phelps, E.B. STREAM SANITATION. New York: John Wiley & Sons,
      1944. 287 p.

      The author totally immerses himself in the life history of streams.
      From the mountain brook to the ocean, streams are things of en-
      ergy, movement, and change. The author describes clearly the
      biochemical reactions and lays the groundwork for the oxygen
      balance, including the mathematical formula that today bears
      his name. As a sanitarian, he puts "natural purification" and
      artificial methods into perspective. For water purity, "perfec-
      tion is unattainable, and 'safe' becomes a purely relative term."
      This classical book is capped with a comprehensive chapter on
      stream microbiology. (For general review, see Appendix A.)

Stream Sanitation—Additional References

422a Klein, L. RIVER POLLUTION II. CAUSES AND EFFECTS. London:
      Butterworth & Co., 1962. 456 p.

422b _____. RIVER POLLUTION III. CONTROL. Washington, D.C.;
      Butterworth & Co., 1966. 484 p.

422c Krenkel, P.A., and Parker, F.L., eds. BIOLOGICAL ASPECTS OF
      THERMAL POLLUTION. Nashville, Tenn.: Vanderbilt University Press,
      1969. 407 p.

a31   Velz, C.J. APPLIED STREAM SANITATION. New York: Wiley-
      Interscience, 1970. 619 p.

Ocean and Estuary Disposal

423   Burchett, M.E.; Tchobanoglous, G.; and Burdoin, A.J. "A Practical
      Approach to Submarine Outfall Calculations." PUBLIC WORKS 98(May
      1967): 95-101.

      The authors have reduced the theoretical approach to dilution,
      dispersion, and decay to a highly useful method for the design
      of submarine outfalls. Practical considerations in outfall calcu-
      lations and diffuser design are delineated. A design procedure
      is formulated and the needed data on receiving waters are iden-
      tified. Sample calculations are included in an example com-
      paring the designs and system costs for four alternatives employ-
      ing ocean discharge of treated effluent.

424   McCarty, P.L., ed. THE NATIONAL SYMPOSIUM ON ESTUARINE
      POLLUTION. Stanford, Calif.: Stanford University, August 1967.
      850 p.

      The proceedings of this symposium contains thirty scientific pa-
      pers dealing with a wide range of estuarine pollution problems.
      Beginning with management and economics of estuarine pollution
      control and proceeding through effects on biota, nutrient effects,
      modeling, and virus and bacteria problems, a good deal of in-
      formation is presented on the San Francisco Bay Estuary. Many
      of the papers are more general in scope, however, especially
      those on trace metal accumulations and pesticide residues.

425   Pearson, E.A., ed. WASTE DISPOSAL IN THE MARINE ENVIRON-
      MENT: PROCEEDINGS OF THE FIRST INTERNATIONAL CONFERENCE.
      New York: Pergamon Press, 1960. 569 p. Paperbound.

      One of the first of its type, this book contains thirty papers
      dealing with wastewater disposal in oceans and estuaries. Sub-

jects range from engineering design considerations to effects of wastewater discharges on marine biota. The risk of infection through bathing in polluted waters was largely discounted as a public health problem. Emphasis is placed on the physics of nearshore oceanography and the methodology of tracer studies. In addition, papers are included on estuarine problems ranging from diurnal oxygen curves to hydraulic and mathematical modeling.

C. EFFLUENT DISPOSAL ON THE LAND

The overall objective of land disposal of effluent is assimilation (return to the environment). Properly engineered systems must be designed to retain wastewater constituents in the soil matrix while avoiding health hazards, groundwater contamination, and other negative environmental impacts.

Documentation

* Engineering of application systems
* Qualitative description of discharge systems
* Effect of constituents on crops
* Generalized design criteria and application procedure
* Cropping practices

Unresolved Issues

* Degree of pretreatment required
* Groundwater effects
* Long-term change in soil composition
* Significance of virus movement
* Aerosol contamination by spraying

General

a16 McGauhey, P.H. ENGINEERING MANAGEMENT OF WATER QUALI-
TY. New York: McGraw-Hill, 1968. Pp. 188-209.

> The engineering aspects of soil treatment systems, including septic-tank leach fields and percolation beds, are detailed. Many important criteria are listed; and the problems of groundwater pollution are addressed. (For general review, see Appendix A.) (Also cited in Section 4, B.)

426 McGauhey, P.H., and Winneberger, J.H. A STUDY OF METHODS OF PREVENTING FAILURE OF SEPTIC-TANK PERCOLATION SYSTEMS. SERL Report no. 65-17. Berkeley: University of California, October

1965. 33 p. Paperbound.

Any soil continuously inundated will lose most of its initial in-
filtrative capacity. In leaching systems, this leads to failure
if the system is designed on the basis of initial infiltration rates.
Several criteria are given for prevention of failure of leaching
systems, including maintenance of aerobic conditions. Such
criteria are also applicable to large scale infiltration systems.

427 Schraufnagel, F.H. "Ridge-and-Furrow Irrigation for Industrial Wastes
Disposal.: JOUR. WPCF 34(November 1962): 1117-32.

The author surveyed the ridge-and-furrow irrigation practice used
by many industries (in Wisconsin) for liquid waste disposal. The
industrial wastes investigated included cannery, dairy, poultry,
tannery, and chemical wastes. Application rates, design features, type of vegetation, and available costs are documented.
A few municipal installations are also described. The author
found that there was such a large spread in liquid loadings on
different soils that basing a proposed system on a single existing
one could easily result in over- or underdesigning by a factor of
ten. The use of drain tiling and the important characteristics
of irrigation water quality are discussed.

428 Sepp, E. "Disposal of Domestic Wastewater by Hillside Sprays." JOUR.
ENV. ENGR. DIV. ASCE 99(April 1973): 109-21.

Findings from a survey of hillside spray installations in California are summarized. Design guidelines put forward by the Bureau of Sanitary Engineering of the California Department of
Public Health are also included. Each system component is discussed, with examples given of successful design criteria. For
each existing system, factors such as flowrate, collection system,
pretreatment, storage facilities, and operations reliability are
discussed. Criteria for spray irrigation design for forested hill-
sides are tabulated.

Effluent Disposal on the Land—Additional References

424a Rafter, G.W. SEWAGE IRRIGATION. U.S. Geological Survey Water
Supply Paper no. 3. Washington, D.C.: U.S. Geological Survey,
1897. 100 p.

428b Schwartz, W.A., and Bendixen, T.W. "Soil Systems for Liquid Waste
Treatment and Disposal: Environmental Factors." JOUR. WPCF 42(April
1970): 624-30.

428c Thomas, R.E. "Land Disposal II: An Overview of Treatment Methods."
JOUR. WPCF 45(July 1973): 1476-84.

Section 5

SLUDGE TREATMENT AND DISPOSAL

Section 5

SLUDGE TREATMENT AND DISPOSAL

Although treatment and disposal is now a critical aspect of any wastewater management program, this area often does not receive adequate consideration in planning and design. In this section, the various aspects of sludge treatment and disposal are covered in the same sequence as they would appear in a treatment scheme. Specific topics considered include: (1) sludge characteristics, (2) sludge thickening and conditioning, (3) sludge treatment, (4) sludge dewatering and drying, (5) sludge incineration, (6) sludge disposal, and (7) sludge recycling.

Disposal of sludge, particularly in large urban areas, is becoming a problem of considerable magnitude. Current interest lies in developing beneficial methods of sludge disposal, such as soil reclamation and feed supplements. The long-term effects of sludge application on land are now receiving considerable attention.

A. GENERAL

501 Burd, R.S. A STUDY OF SLUDGE HANDLING AND DISPOSAL. FWPCA Water Pollution Control Research Series, no. WP-20-4. Cincinnati: FWPCA, May 1968. 369 p. Paperbound. Sold by Government Printing Office.

> Documents various methods of sludge handling and disposal, with emphasis on theoretical development, application examples, and economics. The material is presented in the same sequence as solids are processed in treatment plants, beginning with the grit chamber and ending with ultimate sludge disposal. In all, some twenty operations and processes are discussed separately, with a detailed reference list for each subject. A summary of relative costs for various methods is given. Some ideas for future approaches to separating the solid portion of wastewaters and to treating it more effectively are included. (Also cited in Section 1, B.)

a12 Gloyna, E.F., and Eckenfelder, W.W., Jr., eds. WATER QUALITY IMPROVEMENT BY PHYSICAL AND CHEMICAL PROCESSES. Austin:

Sludge Treatment and Disposal

University of Texas Press, 1970. Pp. 341-436.

Seven papers on various aspects of sludge treatment and disposal are presented. Sludge characteristics, thickening, filtration, centrifugation, and disposal are thoroughly discussed. In a most enlightening paper, the practices of sludge digestion and disposal in Europe are described. (For general review, see Appendix A.) (Also cited in Section 4, A.)

a21 Metcalf & Eddy, Inc. WASTEWATER ENGINEERING: COLLECTION, TREATMENT, DISPOSAL. San Francisco: McGraw-Hill, 1972. Pp. 575-631.

A separate chapter is devoted to the basic practical aspects of sludge handling, treatment, and disposal. (For general review, see Appendix A.) (Also cited in Section 2, A; Section 3, A, B, C, E, F, and H; and Section 4, B.)

a32 Weber, W.J., Jr. PHYSICOCHEMICAL PROCESSES FOR WATER QUALITY CONTROL. New York: John Wiley & Sons, 1972. Pp. 533-96.

This comprehensive coverage of sludge treatment and disposal was written by Richard Dick, an acknowledged expert on the subject. In addition to descriptions of all the unit processes and operations in sludge treatment, the costs of technology and the feasible alternatives for sludge management are considered. The thorough literature review on each subject is another useful aspect of this presentation. (For general review, see Appendix A.) (Also cited in Section 3, B, E, F, and H.)

B. SLUDGE CHARACTERISTICS

The residual solids resulting from various wastewater treatment processes are referred to as sludge. Sludge can be characterized by its chemical content, physical properties, and biological components. The source of the sludge will affect its ultimate treatment and disposal.

Documentation

* Generalized composition of sludge by type
* Expected heavy metal contents
* Odor-producing substances

Unresolved Issues

* Prediction of toxic substances
* Heavy metals' effects when applied to the land

502   Kaplovsky, A.J., and Genetelli, E. "Sludge Characteristics of Municipal Solids." In PROCEEDINGS OF THE SYMPOSIUM ON LAND DISPOSAL OF MUNICIPAL EFFLUENTS AND SLUDGES, pp. 3-18. EPA-902/9-73-001. New York: EPA, March 1973. Paperbound.

The practice of land spreading of municipal sludge is reviewed in this paper. Sludge conditioning techniques are given along with detailed tabulations of the composition of various sludges. In the future, the method of land application that is practiced should dictate the acceptable characteristics of the sludge to be applied.

503   Rains, B.A.; DePrimo, M.J.; and Groseclose, I.L. ODORS EMITTED FROM RAW AND DIGESTED SEWAGE SLUDGE. Environmental Protection Technology Series, EPA-670/2-73-098. Washington, D.C.: EPA, December 1973. 69 p. Paperbound. Sold by Government Printing Office.

Gases evolved from raw and digested sludge were sampled and characterized, and the organic chemicals responsible for odors were identified. Odor control methods were studied, including air dilution, activated carbon adsorption (not completely effective), and chlorine oxidation. The threshold levels and characteristics of the identified odors were analyzed in this well-documented and illustrated study.

C. SLUDGE THICKENING AND CONDITIONING

Thickening and conditioning of sludge involves operations and processes that can be used to reduce the moisture content and prepare the sludge for further treatment. The thickening operations considered include gravity and dissolved-air flotation thickening, while conditioning includes elutriation, freezing, irradiation, chemical addition, and heat treatment.

Documentation

*   Design of conventional operations and processes
*   Expected performance from conventional processes

Unresolved Issues

*   Prediction of effluent quality
*   Degree of conditioning required for subsequent processing
*   Effects of combining primary and secondary sludges

504   Everett, J.G. "Dewatering of Wastewater Sludge by Heat Treatment." JOUR. WPCF 44(January 1972): 92-100.

Heat treatment of sludge involves heating to temperatures of 150°
to 220° C. in the absence of air. The effects of detention time
and temperature on the dewatering characteristics of different
sludges were investigated. The solubilization of suspended organics was also tested, and it was found to vary directly with
the temperature for each sludge studied. For all sludges, heat
treatment improved their dewatering characteristics by destroying
the structural integrity of microorganisms, thereby preventing them
from packing together. High temperatures and longer holding
times yield more cell destruction and improve dewatering characteristics.

505 Katz, W.J., and Geinopolos, A. "Sludge Thickening by Dissolved-Air
Flotation." JOUR. WPCF 39(June 1967): 946-57.

In dissolved air flotation, thickening is accomplished by dissolv-
ing air in water at high pressure and then reducing the pressure,
thus forming small air bubbles that attach themselves to sludge
particles and float them to the surface, where the sludge is re-
moved. Results of testing for final sludge concentration and
percentage of solids recovery at six plants throughout the country are presented. The design configuration and operating performance of each system are included. Addition of polymers resulted in 99 percent solids recovery and concentrations of 4 percent and more. Chemical dosage was related to solids loadings
and thickened sludge concentration, and chemical costs were
computed.

D. SLUDGE TREATMENT

The biological treatment of sludge involves the reduction of the volatile portion
of the residual solids. This is normally accomplished either aerobically (in the
presence of oxygen) or anaerobically (in the absence of oxygen).

Documentation

*   Qualitative description of processes
*   Gas production from anaerobic decomposition
*   Basic design data and criteria

Unresolved Issues

*   Effects of toxic constituents
*   Optimum process application
*   Expected effects on constituents
*   Optimum reuse of treatment by-products

Aerobic Digestion

506   Ahlberg, N.R., and Boyko, B.I. "Evaluation and Design of Aerobic Digesters." JOUR. WPCF 44(April 1972): 634-43.

> The authors conducted an eighteen-month study of the performance and design parameters for seven aerobic digesters in Ontario, Canada. All the key design factors, including minimum acceptable values and their variations with temperature, are reported. A great deal of useful data is contained in this well-organized article.

507   Ritter, L.E. "Design and Operating Experiences Using Diffused Aeration for Sludge Digestion." JOUR. WPCF 42(October 1970): 1782-91.

> The operation, process description, and problems of aerobic digestion of waste-activated sludge are reported. Three plants in Pennsylvania were chosen for detailed investigation because of design variations and percentages of industrial wastes. Design and operating data are presented; and specific conclusions relating to satisfactory operating procedures, temperature ranges, process modifications, and resulting costs are given.

Anaerobic Digestion

508   McCarty, P.L. "Anaerobic Waste Treatment Fundamentals." PUBLIC WORKS 95(September-December 1964).

> This four-part series of articles is perhaps the most complete presentation available on the fundamentals of anaerobic sludge digestion. In part 1, the chemistry and microbiology are delineated, along with a discussion of the advantages and disadvantages of anaerobic treatment. In part 2, the environmental requirements, optimum temperatures, lack of oxygen, nutrient levels, absence of toxic materials, and optimum pH range, are given. A useful chart is given in which the pH and alkalinity are related and normal operational limits specified. Causes and controls of treatment imbalance are discussed. In part 3, toxicities of common ions are explored with concentration ranges given for stimulatory, moderately inhibitory, and strongly inhibitory conditions. In part 4, the application of various treatment concepts to plant design is reviewed.

509   Water Pollution Control Federation. ANAEROBIC SLUDGE DIGESTION. Manual of Practice no. 16. Washington, D.C.: 1968. 74 p. Paperbound.

> Designed for the wastewater treatment plant operator, this manual includes common analytical tests, start-up procedures, and digester cleaning practice. Characteristics of sludge that affect its digestion, a brief synopsis of biological considerations, and

some design criteria are included. The manual is filled with photographs of equipment and diagrams that serve to clarify subjects such as gas-lift mixing. Warnings are also included against miraculous "cures" to digestion problems, such as the use of packaged enzymes.

Sludge Lagoons

510 Dornbush, J.N. "State of the Art: Anaerobic Lagoons." In SECOND INTERNATIONAL SYMPOSIUM FOR WASTE TREATMENT LAGOONS, pp. 382-87. Edited by R.E. McKinney. Lawrence: University of Kansas, June 1970.

The operative treatment mechanisms for anaerobic lagoons, whether for treating waste sludge, livestock wastes, or raw wastewater, are reviewed. Emphasis is on successful operating conditions; attention to proper temperature, pH, and mixing; and complete design criteria. The need to clearly define thermal requirements, loading and detention times for optimum treatment, and the relative importance of sludge accumulations as the major source of nuisance odors are delineated as potential research areas.

511 Parker, C.D., and Skerry, G.P. "Function of Solids in Anaerobic Lagoon Treatment of Wastewater." JOUR. WPCF 40(February 1968): pt. 1, 192-204.

A number of anaerobic lagoons in Australia were examined in field studies. The purpose of the study was to assess gas production and solids activity under varying field conditions. Surprisingly, the presence of algae in these relatively shallow ponds did not hinder the anaerobic decomposition. Mixing of actively decaying solids enchanced the reduction of organic matter.

Other Methods

512 Farrell, J.B., et al. "Lime Stabilization of Primary Sludges." JOUR. WPCF 46(January 1974): 113-22.

Ordinarily, sludge is biologically stabilized (or conditioned) by anaerobic or aerobic digestion; however, chemical stabilization using chlorine or lime is another option. EPA research with liming of primary sludge prior to dewatering and disposal is reported. Microbiological investigations were conducted, and it was found that an immediate, but not long-term, bacterial kill was effected. Liming reduced the concentration of solids in the primary sludge, but increased the vacuum filter yields.

E. SLUDGE DEWATERING AND DRYING

Dewatering methods are used to further reduce the moisture content of thick-ened, treated, or conditioned sludge. Dewatering can be accomplished by spreading the sludge on sand beds or by mechanical means such as vacuum or pressure filters and centrifuges.

Documentation

* Qualitative process description
* Design criteria
* Economic advantages

Unresolved Issues

* Power input
* Degree of dryness required
* Effect of preconditioning and treatment
* Long-term reliability
* Optimum process application

513 O'Donnell, C., and Keith, F., Jr. "Centrifugal Dewatering of Aerobic Waste Sludges." JOUR. WPCF 44(November 1972): 2162-71.

Upflow solid-bowl centrifuges have been developed to capture solids in waste-activated sludge very efficiently. The centri-fuge is fed continuously except for brief periods during which cake removal is accomplished automatically. Suspended solids capture of 90 to 95 percent without chemical addition is prac-tical. Cake concentrations range from 5 percent solids for industrial sludge to 15 percent for sludge from the Unox pure oxygen-activated sludge process. Aerobically digested sludge can be concentrated to nearly 10 percent in a residence time of ten minutes. Alum sludges from water treatment plants, which are usually very difficult to dewater, have been dewa-tered successfully with this imperforate bowl centrifuge.

514 Parkhurst, J.D., et al. "Dewatering Digested Primary Sludge." JOUR. WPCF 46(March 1974): 468-84.

Several sludge dewatering alternatives were evaluated by the county sanitation districts of Los Angeles. Based on the results of pilot plant studies, it was found that five conditioning-dewatering schemes were capable of meeting the requirement of 95 percent cap-ture of solids from anaerobically digested primary sludge. Although alternatives involving polymers, heat, and chemicals as condition-ing agents in conjunction with centrifuges and vacuum and pressure filters met the criteria, a two-stage centrifugation system, using hor-izontal scroll centrifuges (existing) followed by basket centrifuges, was selected. Cost estimates are given for all five alternatives.

515 Water Pollution Control Federation. SLUDGE DEWATERING. Manual of Practice no. 20. Washington, D.C.: 1969. 115 p. Paperbound.

A summary of current information on design and operating practices for reducing the moisture content of sludge is presented in this manual. Expanded from the original subject of vacuum filtration, this manual also includes dewatering by centrifugation and gravity, in special cells, and on sand beds. The theory of chemical conditioning is explained, and filterability tests such as the Buchner funnel method, the filter leaf method, and the specific resistance method, are described. This is an invaluable manual for design of sludge dewatering facilities as it contains specific design criteria and sample specifications.

F. SLUDGE INCINERATION

Incineration is a natural extension of the dewatering process and is used to convert dewatered sludge into an inert ash. For relatively dry sludge, the process usually can be sustained without supplemental fuel. The stack gases may require treatment, and the inert ash must be disposed of or utilized.

Documentation

* Physical configuration of conventional processes
* Process design data and criteria
* Energy requirements

Unresolved Issues

* Air pollution effects
* Optimum process application

516 Balakrishnan, S.; Williamson, D.E.; and Okey, R.W. STATE OF THE ART REVIEW OF SLUDGE INCINERATION PRACTICE. FWQA Water Pollution Control Research Series, 17070 DIV. Washington, D.C.: FWQA, 1970. 135 p. Paperbound. Sold by Government Printing Office.

Analyzes the present and future applications of this process, going beyond the engineering and cost considerations associated with sludge incineration. Attitudes of consulting engineers were surveyed, and it was found that incineration was preferred for urban populations of more than 15,000. Operational aspects are discussed briefly, along with an analysis of incineration of materials other than dewatered municipal sludges. Pricing information is included for the major brands of incinerators.

517   Environmental Protection Agency Task Force. SEWAGE SLUDGE INCIN-
      ERATION. EPA-R2-72-060. Washington, D.C.: EPA, August 1972.
      90 p. Paperbound. Sold by Government Printing Office.

      A task force was established within the EPA to evaluate sludge
      incineration as a sludge disposal alternative. The emphasis was
      on the emissions of pollutants such as polychlorinated biphenyls
      (PCBs) and pesticides. Considerable data are appended to sup-
      port the conclusion that such materials are destroyed during
      sludge incineration. Incineration as an environmentally sound
      alternative (energy considerations notwithstanding) to ocean dis-
      posal is strongly supported by the conclusions presented in this
      report. Because insufficient data are available on health ef-
      fects of atmospheric emissions from incinerators, a plea is made
      for the collection of such information.

G. SLUDGE DISPOSAL

Methods of ultimate sludge disposal include burial in landfills, controlled dumping into abandoned mines and quarries, discharging to the ocean, and spreading or incorporating into soil. Land application and ocean disposal of sludge
have created emotional controversies over potential ecological damage and
health hazards.

Documentation

*   Qualitative description of application methods
*   Design criteria

Unresolved Issues

*   Long-term effects of sludge disposal practice
*   Alternative methods of sludge transport
*   Long-term effects on soil composition
*   Health considerations
*   Legality of disposal practices

Land Disposal

518   Ewing, B.B., and Dick, R.I. "Disposal of Sludge on Land." In WATER
      QUALITY IMPROVEMENT BY PHYSICAL AND CHEMICAL PROCESSES,
      pp. 394-408. Edited by E.F. Gloyna and W.W. Eckenfelder, Jr. Aus-
      tin: University of Texas Press, 1970.

      This paper contains an excellent synopsis of the literature, prob-
      lems, and issues associated with land disposal, including such
      topic areas as sludge composition, disease transmission, nutrients,

heavy metals, and nuisances. Cost curves are presented in which the major variable is the distance in miles to the disposal point. For large cities, pipeline transport is most economical, with land disposal being more economical than landfill of dewatered sludge up to a haul of about thirty-five miles.

519   Walker, J.M. "Sludge Disposal Studies at Beltsville." In PROCEEDINGS OF THE SYMPOSIUM ON LAND DISPOSAL OF MUNICIPAL EFFLUENTS AND SLUDGES, pp. 102-16. EPA-902/9-73-001. New York: EPA, March 1973. Paperbound.

The results of four different field studies involving the Agricultural Research Service, EPA, and the state of Maryland are reported. With 600 tons of 20 percent sludge to dispose of daily from the Blue Plains sewage treatment plant, the limits of sludge application to the land had to be probed. Digested sludge was plowed, disked, and rototilled into the surface layer, and crops were grown and analyzed for heavy metals. Raw sludge was incorporated into the soil by trenching at rates up to 500 tons per acre. Finally, large scale composting, initiated in September 1972, was also found to be successful.

Ocean Disposal

520   Domenowske, R.S., and Matsuda, R.I. "Sludge Disposal and the Marine Environment." JOUR. WPCF 41(September 1969): 1613-24.

A good deal of emotion surrounds the subject of sludge disposal through ocean outfalls. In this paper, a fairly objective view is presented of the situation at Seattle, Washington, based on extensive sampling and the taking of underwater photographs. Although there was considerable fluctuation in the observations, the biological fauna in the immediate vicinity of the outfall appeared to be normal in both numbers and diversity. The observed diffuser plumes appeared to be functioning exactly as predicted in model studies.

521   Heckroth, C.W. "Special Report: Ocean Disposal - Good or Bad?" WATER AND WASTES ENGINEERING 10(October 1973): 32-38.

Ocean disposal of both effluents and sludges is analyzed in terms of environmental damage, current criteria, and governmental regulations. Facts as well as the opinions of well-known ecologists are cited. The findings of a detailed three-year study of the problem, the Southern California Coastal Water Research Project report, are presented with a comparison of the total amounts of pollutants discharged to the ocean from municipal wastewater, industrial wastewater, and surface runoff.

522   Minges, M.C. "Ocean Disposal Subject to New Controls." JOUR. WPCF 45(May 1973): 782-83.

By requiring a permit to discharge wastes to the ocean, Congress has closed a potential environmental loophole in Public Law 92-500, the 1972 Federal Water Pollution Control Act Amendments. The history and effects of ocean dumping are cited briefly, along with the adverse impact on humans of uncontrolled dumping. The act is reviewed with respect to permit requirements and penalties and types of pollutants.

H. SLUDGE RECYCLING

One of the exciting prospects for new technological development is the reuse of waste solids. Soil conditioning, soil reclamation, and animal feed supplements are among the possible uses for sludge. Many details need to be delineated for the latter use. The application of sludge for soil conditioning and fertilizing is well documented.

Documentation

* Qualitative description of applications and potential uses
* Generalized design data

Unresolved Issues

* Optimum process application
* Pretreatment requirements
* Long-term effects of trace elements
* Health considerations
* Application rates

Soil Conditioning

523 Dotson, G.K. "Some Constraints of Spreading Sewage Sludge on Cropland." In PROCEEDINGS OF THE SYMPOSIUM ON LAND DISPOSAL OF MUNICIPAL EFFLUENTS AND SLUDGES, pp. 67-79. EPA-902/9-73-001. New York: EPA, March 1973. Paperbound.

Objectives in land spreading of sludge include: (1) disposal only, (2) reclamation of unproductive soils, or (3) the addition of fertilizer, water, and organic matter to cropland. The constraints discussed in this paper deal with the latter two objectives. The major constraints associated with sludge composition are nitrogen, heavy metals, and pathogens. The state of research in these areas is given. Soil properties, legal restraints, and public resistance may dictate the type of sludge to be spread.

524   Page, A.L. FATE AND EFFECTS OF TRACE ELEMENTS IN SEWAGE SLUDGE WHEN APPLIED TO AGRICULTURAL LANDS: A LITERATURE REVIEW STUDY. Environmental Protection Technology Series, EPA-670/2-74-005. Washington, D.C.: EPA, January 1974. 96 p. Paperbound. Sold by Government Printing Office.

> Concentrations of trace elements in sewage sludges are related to industrial and consumer uses and vary widely between communities. Ranges of each element reported for sludges from approximately 300 treatment plants in the United States and Europe are given. Where concentrations exceed these ranges for one or more trace elements, a significant industrial input is likely. Exceptions to this are copper and zinc, which often are contributed by contamination from metal pipes and tanks during water storage, treatment, or conveyance. Yields of crops receiving various loadings are reviewed and discussed. Potential health problems and future research areas are outlined.

525   Water Pollution Control Federation. UTILIZATION OF MUNICIPAL WASTEWATER SLUDGE. Manual of Practice no. 2. Washington, D.C.: 1971. 47 p. Paperbound.

> Uses for treated sludge include soil conditioning, soil fertilizing, soil reclamation, and landfill. Because modern commercial fertilizers contain 20 to 30 percent nitrogen as opposed to 2 to 6 percent for sludge, sludge is most often used as a supplement rather than a replacement for fertilizer. Processes from which a sterilized soil conditioner is produced include heat treatment and composting. Heating sludge to 135° F. for one hour should precede its use as a general soil amendment. The real problems associated with marketing sludge as a soil conditioner limit this disposal method.

Soil Reclamation

526   Dalton, F.E.; Stein, J.E.; and Lynam, B.T. "Land Reclamation: A Complete Solution of the Sludge and Solids Disposal Problem." JOUR. WPCF 40(May 1968): pt. 1, 789-804.

> The problems and need for sludge disposal in Chicago are described. A long-range plan involving the application of solids on the land is envisioned in which 30,000 acres of land in southern Illinois would be used. The only novel aspect of this plan is its size. Other large facilities around the world are documented in a very interesting appendix. The advantages and costs of land application are described, along with the current status of model facilities at the Hanover Treatment Plant.

527   Kudrna, F.L., and Kelley, G.T. "Implementing the Chicago Prairie Plan." In RECYCLING TREATED MUNICIPAL WASTEWATER AND SLUDGE THROUGH FOREST AND CROPLAND, pp. 364-70. Edited by

W.E. Sopper and L.T. Kardos. University Park: The Pennsylvania State University Press, 1973.

This article concerns the much-heralded Chicago Prairie Plan in Fulton County, Illinois. Some 7,000 acres of strip-mined land about 200 miles southwest of Chicago were purchased as part of a sludge disposal-land reclamation project. Transportation costs, additional uses of the purchased land, and future plans are discussed.

528   Lejcher, T.R., and Junkel, S.A. "Restoration of Acid Spoil Banks with Treated Sewage Sludge." In RECYCLING TREATED MUNICIPAL WASTE-WATER AND SLUDGE THROUGH FOREST AND CROPLAND, pp. 184-99. Edited by W.E. Sopper and L.T. Kardos. University Park: The Pennsylvania State University Press, 1973.

The preliminary results of a reclamation demonstration project in a southern Illinois strip-mined area are detailed in this paper. On the basis of initial observations, it appears that treated municipal sludge, when applied to the spoil in sufficient amounts, improves spoil pH, allows establishment of vegetation, and reduces acidity in the runoff.

Feed Supplements

529   Lin, S.C., and Witherow, J.L. "Cattle Paunch Contents as Fish Feed Supplement: Feasibility Studies." In PROCEEDINGS OF THE THIRD NATIONAL SYMPOSIUM ON FOOD PROCESSING WASTES, pp. 401-8. Environmental Protection Technology Series, EPA-R2-72-018. Corvallis, Oreg.: EPA, March 1972. Paperbound. Sold by Government Printing Office.

The use of dried cattle rumen material as supplemental feed in the rapidly growing catfish farming industry is discussed. Converting this potentially polluting abattoir waste into a useful product requires screening and air drying prior to direct feeding to catfish. Despite the fact that well-documented studies on parasite and bacterial transmission are necessary, it is concluded that the conversion of waste protein into useful forms is promising.

530   Young, L.J., and Purcupile, J.C. "Food from Sewage? It Can Be Done." WATER AND WASTES ENGINEERING 10(June 1973): 47-50.

As documented in this paper, yeast can be grown economically on sterilized supernatant from sludge treatment processes. From an economic analysis standpoint, the system employing heat treatment, yeast production, and incineration showed a profit on an annual cost basis. The authors conclude that technical and market development are needed to confirm or deny the assumptions underlying this apparent economic justification.

Section 6

ECONOMIC CONSIDERATIONS

Section 6

ECONOMIC CONSIDERATIONS

Economic considerations in wastewater management involve: (1) assessment of alternative plans on an equitable basis, (2) application of basic cost data, and (3) economic planning encompassing financing and repayment. The use of new or alternative technology depends heavily on the cost of that technology and the resulting economic benefits that can be quantified.

The practical application of economics to wastewater management has been the delineation of the costs of unit operations and processes. More recently, cost sharing, industrial user charges, and federal and state grant programs for construction of new facilities have attracted attention. From a theoretical standpoint, subjects such as cost effectiveness and time-capacity expansion have been of interest. The application of these approaches to economic planning should become more widely accepted as funds for construction become limited.

A. ASSESSMENT OF ALTERNATIVES

Both economic and financial analyses are necessary for an assessment of wastewater management alternatives. An economic analysis is used to assess the wisdom of and need for investment of valuable resources. The financial analysis is conducted to determine the cost burden that will be placed upon individuals and industries located within the project area.

Documentation

*   Methods of performing cost comparisons using present worth or annual cost
*   Standardized procedures published by the federal government

Unresolved Issues

*   Evaluation of social benefits and costs
*   Evaluation of environmental benefits and costs
*   Sensitivity of design criteria to costs

Economic Considerations

* How to deal with inflation in economic analyses
* Application of time-capacity models in facilities planning
* Appropriate interest rate and return period for cost analysis

Economic Criteria

601 Case, F.H. "Economics of Water Quality and Wastewater Control." JOUR. SAN. ENGR. DIV. ASCE 98(April 1972): 427-34.

The author takes the position that water and wastewater are the same economic goods, subject to the same concepts, and must be treated as a single economic good. As water production becomes more costly and difficult, water production planning should include considerations of: (1) socioeconomic problems to be solved, (2) contributions of water to the solution of socioeconomic problems, (3) the costs and net benefits to particular groups and to society associated with each solution, (4) the willingness and capacity of each group to accept various cost-benefit trade-offs, and (5) the solutions-matrices associated with each problem.

602 Sliter, J.T. "Is There a Solution to Construction Cost Escalation?" JOUR. WPCF 46(September 1974): 2082-85.

The problem of inflation and cost indexing as it relates to wastewater treatment plant construction costs is discussed. Caught between rising costs and fixed grants, plants currently under construction are being delayed and future plant construction postponed. Some of the causes and effects of inflation are examined, and the points of view of industry, municipalities, and the EPA are documented.

Cost-Effectiveness

603 English, J.M., ed. COST-EFFECTIVENESS: THE ECONOMIC EVALUATION OF ENGINEERED SYSTEMS. New York: John Wiley & Sons, 1968. 301 p.

Although it is aimed at the entire spectrum of engineering, with emphasis on aerospace and military engineering, considerable insight into the discipline of cost-effectiveness can be gained from this book. Fundamentally, it is nothing more than basic engineering economics. The process of combining costs (expenses) and effectiveness (benefits) can produce many fallacies, which are explored in a separate chapter. The book was produced from notes prepared by the seven authors for a one-week short course offered in April 1967 at UCLA.

604 Ko, S.C., and Duckstein, L. "Cost-Effectiveness Analysis of Wastewater Reuses." JOUR. SAN. ENGR. DIV. ASCE 98(December 1972): 869-81.

The Kazanowski approach to cost-effectiveness analysis, using the practical example of wastewater reuse alternatives at Tucson, Arizona, is discussed. The ten steps in the cost-effectiveness method are delineated and illustrated. Latent benefits and costs are quantified, and sensitivity analysis is conducted on the major cost variables.

Time-Capacity Expansion

605 Rachford, T.M.; Scarato, R.F.; and Tchobanoglous, G. "Time-Capacity Expansion of Waste Treatment Systems." JOUR. SAN. ENGR. DIV. ASCE 95(December 1969): 1063-77.

A method of evaluating the economic efficiency of a planned project involving the phased expansion of wastewater treatment works is presented and illustrated. The major parameters are the economy-of-scale factor and the time factor. The optimal period between expansion phases for various interest rates and economy-of-scale factors can be determined. The method is applicable to components as well as to entire treatment and collection systems. The mathematical model developed originally by Manne is explained, and useful diagrams are included for determination of the optimum time and size of plant expansion.

B. BASIC COST DATA

Basic cost data for the construction of collection, treatment, and disposal facilities are derived from bid and actual construction cost information. Generalized cost data, in which the cost of a treatment process is related to the average flowrate, are often used in estimating costs in planning studies. To be useful at different times and locations, cost data must be related to construction cost indexes, wholesale price indexes, or wage rates.

Documentation

* Generalized cost information for unit operations and processes
* Indexes for updating construction costs
* General methodology of presenting costs

Unresolved Issues

* Standardized cost data for individual treatment components
* Adequate weighting of indexes to reflect treatment process differences
* Appropriate indexes for rural versus urban construction

606 Blecker, H.G., and Cadman, T.W. CAPITAL AND OPERATING COSTS

OF POLLUTION CONTROL EQUIPMENT MODULES. 2 vols. Socioeconomic Environmental Studies Series, EPA-R5-73-0236. Washington, D.C.: EPA, July 1973. Vol. 1, 255 p.; vol. 2, 183 p. Paperbound. Sold by Government Printing Office.

Basic cost data for some eighty-eight separate components or modules of pollution control flowsheets are contained in this two-volume manual. Components included in pollution control flowsheets can be used for municipal and industrial wastes that are solid, liquid, and gaseous. To illustrate the complex and detailed evaluation procedures, user information and sample worksheets are included in volume 1. Cost curves for each component, ranging from air compressors to steam boilers, are given in volume 2. Although too detailed for conceptual planning, the cost data would be most useful to design engineers.

607   Nesbitt, J.B. "Cost of Spray Irrigation for Wastewater Renovation." In RECYCLING TREATED MUNICIPAL WASTEWATER AND SLUDGE THROUGH FOREST AND CROPLAND, pp. 334-38. Edited by W.E. Sopper and L.T. Kardos. University Park: The Pennsylvania State University Press, 1973.

The costs of hypothetical spray irrigation systems with capacities of one, five, and ten mgd are estimated. Annual costs (amortized capital costs plus operating costs) are 12.7 cents, 8.0 cents, and 7.2 cents per thousand gallons for each of the three flowrates respectively. Charts and assumptions are given.

608   Patterson, W.L., and Banker, R.F. ESTIMATING COSTS AND MANPOWER REQUIREMENTS FOR CONVENTIONAL WASTEWATER TREATMENT FACILITIES. EPA Water Pollution Control Research Series, 17090 DAN. Washington, D.C.: EPA, October 1971. 251 p. Paperbound. Sold by Government Printing Office.

Basic cost data for preliminary engineering cost estimates are presented in this comprehensive report. Costs are related to the relevant variables for each unit operation and process. Planners and nonengineers wishing to compare the cost of trickling filters to activated sludge will be thwarted by the lack of simplified curves relating to flowrates and costs. Examples are included to illustrate the use of the curves to estimate construction and operation and maintenance costs, as well as manpower requirements.

609   Smith, R. "Cost of Conventional and Advanced Treatment of Wastewater." JOUR. WPCF 40(September 1968): 1546-74.

This article is one of the first comprehensive efforts at amassing basic cost data for treatment processes. Previous cost reports are analyzed, updated, and compared to the author's cost curves. Seventeen cost curves are presented, along with cost indexes for capital, amortized capital, operation and maintenance, and total costs. An additional fifteen curves are presented for more de-

tailed construction cost estimates for unit processes. This is a
useful and comprehensive source of cost data.

610 Smith, R., and Eilers, R.G. COST TO THE CONSUMER FOR COLLEC-
TION AND TREATMENT OF WASTEWATER. EPA Water Pollution Control Research Series, 17090. Washington, D.C.: EPA, July 1970. 86
p. Paperbound. Sold by Government Printing Office.

Costs of wastewater management on a national scale are presented in dollars per capita. The cost of wastewater management
is compared with the need for collection and treatment facilities
and the cost for other utilities such as water supply. Interest-
ingly, the cost of wastewater collection was found to be three
times as expensive as treatment, based on a 1968 EPA inventory
of facilities. The cost to the consumer of collection and treatment represents about 0.1 percent of the national personal ex-
penditures in 1967, or about $570 million.

C. ECONOMIC PLANNING

Determination of the cost effectiveness of emerging and available technology
is an important aspect of regional and long-range economic planning. Formal-
ized planning procedures and standardized cost-effectiveness analysis guidelines
must be followed to secure governmental grants for the construction of treatment facilities.

Documentation

* Generalized procedures for cost distribution
* Cost of available technology
* Guidelines for federal grant programs
* State planning regulations

Unresolved Issues

* Need for regionalization
* Effect of interest rate on selection of planning period
* Cost of emerging and undeveloped technology
* Allocation and implementation of user charges
* Equitable distribution of costs for sewerage service
* Long-term availability of grant programs

Cost of Technology

611 Monti, R.P., and Silberman, P. "Wastewater System Alternctes: What
Are They...and What Cost?" WATER AND WASTES ENGINEERING

11(March-June 1974).

Treatment and disposal alternatives for meeting secondary treatment, best practicable treatment, and best available treatment are introduced in flowsheets for water-oriented and land-oriented disposal in part 1 of this four-part series of articles. In part 2, the system components and water quality criteria are discussed and cost curves are presented for advanced wastewater treatment processes. The ultimate goals of Public Law 92-500 are discussed, and process flowsheets and cost curves are presented in parts 3 and 4.

612   Roper, W. "Wastewater Management Studies by the Corps of Engineers." JOUR. ENV. ENGR. DIV. ASCE 99(October 1973): 653-69.

The cost of technology necessary to achieve wastewater management on a regional scale has been studied by the U.S. Army Corps of Engineers in five different cases. The study reported dealt with the Chicago area, which has a projected wastewater flowrate of 2,376 mgd. Alternative technologies considered were advanced biological treatment, complete physical-chemical treatment, and land treatment by spray irrigation. Although the Corps did not rate the alternative systems, land treatment was approximately half the cost of the other two technologies. In addition to the cost comparison, the chemical and energy requirements and resultant effluent qualities were compared.

Financing and Repayment

613   American City Magazine. MUNICIPAL SEWER SERVICE CHARGES. New York: Buttenheim Publishing Corp., 1970. 108 p.

Municipal sewer service charge programs in selected areas throughout the United States are presented in this booklet. A brief look is given at the sewerage systems financed with these charges. Written prior to the issuance of federal guidelines, it discusses some interesting innovations and local remedies to the problem of equitable and practical financing of wastewater management. In addition, special districts and intermunicipal sewerage authorities are cited and examples of successful rate structures are given. Examples of resourceful designs that make treatment facilities city assets, such as parks, marinas, and art or community display centers, are also offered.

614   Fish, H. "Pollution Control Financing in the United Kingdom and Europe." JOUR. WPCF 45(April 1973): 734-41.

The author presents a clear and forceful account of the possible, equitable, and effective methods of pollution control financing. The present positions of the United Kingdom, the Netherlands, Germany, and France are reviewed and criticized. Changes that are probable and those that are needed in attaining water pollution control are specified in a lucid and thoughtful manner.

Grant Programs

a09 Environmental Protection Agency. GUIDANCE FOR FACILITIES PLAN-
     NING. 2nd ed. Washington, D.C.: 1974. 115 p. Paperbound.

These guidelines were developed for reports on facilities plans
so that federal grants could be disbursed for projects represent-
ing cost-effective alternatives for wastewater management. A
detailed list of federal regulations and pertinent references precedes the orderly delineation of the steps required in the planning process. The relationship of facilities plans and basin
plans, areawide waste treatment management plans, and municipal permits are explained. A sample outline of a facilities
plan is included. (For general review, see Appendix A.)
(Also cited in Section 1, C and D.)

Section 7

PLANNING AND IMPLEMENTATION

Section 7

PLANNING AND IMPLEMENTATION

Because planning and implementation concepts for wastewater management are closely related, they must be considered together. For example, regional consolidation is meaningless without an implementation program that provides a workable arrangement between institutions and agencies involved in water pollution control. In addition to planning for shifts in population and a more equitable allocation of resources, environmental considerations include definition of environmental quality, conservation of ecosystem diversity, and planning mechanisms to assess environmental impacts.

Basin and areawide planning have become increasingly important, and many case studies of comprehensive planning efforts exist. Unresolved and controversial issues include proper resource allocation, innovative institutional arrangements, and meaningful assessments of the social impact of wastewater management.

A. PLANNING FACTORS

Agreement on the necessary planning arrangements by the agencies and institutions involved is fundamental to the implementation of wastewater management plans. In addition to economics, the allocation of resources for wastewater management and determination of populations to be served are important planning factors.

Documentation

* Generalized planning guides
* Conventional population projection methods
* Basin planning involving multiple agencies
* Alternative types of local and regional planning agencies

Unresolved Issues

* Accuracy and use of population projection procedures

Planning and Implementation

* Local versus regional management
* Proper resource allocation

Population Trends and Resource Use

701 Gray, S.L., and Young, R.A. "The Economic Value of Water for Waste Dilution: Regional Forecasts to 1980." JOUR. WPCF 46(July 1974): 1653-62.

The use of surface water resources primarily for waste dilution is a concept whose time has past. Nevertheless, the authors have estimated the value of dilution water projected to 1980 population conditions for each of twenty-two water resource regions in the United States. The value was found to be marginal in most cases, implying higher uses of the water and illustrating that waste treatment is generally less costly than low flow augmentation as a means for water quality improvement.

702 McJunkin, F.E. "Population Forecasting by Sanitary Engineers." JOUR. SAN. ENGR. DIV. ASCE 90(August 1964): 31-58.

Population projection techniques used commonly by the sanitary engineering profession are evaluated critically. Techniques considered include graphical, mathematical, or formula-type methods; statistical methods of correlation and regression analysis; ratio methods using other estimates; and component methods. This is a comprehensive, if somewhat dated, review of data sources and techniques also used by planners, economists, and demographers. Each method is illustrated, and an extensive bibliography is included.

Regional Consolidation and Institutional Arrangements

703 Cleary, E.J. "Quality Management." JOUR. WPCF 42(February 1970): pt. 1, 157-64.

The author discusses innovations under way on the state level that experiment with more dynamic institutional arrangements than the standard methods of operation, which involve myriads of regulations. Interstate compacts, as in the Delaware River basin, and watershed conservancy districts, as in the Miami River Valley of Ohio, are carefully analyzed. The author believes that, on the basis of the cited examples, the threshold of a productive era of innovation in designing administrative machinery is near at hand.

704 Jacobs, R.L., and Hoge, C.H. "Developing a Regional Wastewater System." JOUR. WPCF 42(November 1970): 1951-61.

Examines wastewater management planning for metropolitan complexes where many contiguous governmental and private entities attempt, with varying degrees of effort and effectiveness, to deal with the problem on an individual basis. In these situations, regional sewerage systems are often the most practical solutions. There are, however, considerations such as engineering, legal and financial problems, industrial wastes, communications, and bonding that must be dealt with before a regional solution can be secured. The concept is illustrated by examining regional systems in Texas.

705 Yao, K.M. "Regionalization and Water Quality Management." JOUR. WPCF 45(March 1973): 407-11.

The three "Rs" of wastewater management planning--regionalism, reliability, and recycling--are discussed. Regionalization is encouraged officially by providing an additional 10 percent in federal grants for projects deemed to be regional in nature. The author explores the effects of regionalization on water quality and concludes that for some cases it may be the necessary step for achieving the desired stream quality objectives.

B. ENVIRONMENTAL CONSIDERATIONS

A major objectives of environmental planning is the analysis of the possible outcome of man's actions on the environment. Preservation of ecological diversity and the minimization of irreversible impacts are important goals.

Documentation

* Generalized approach to the assessment of impacts
* Existing environmental quality

Unresolved Issues

* Response of ecosystems to the various constituents in wastewater effluents
* Assessment of social impact
* Practical definition of environmental quality

Environmental Quality

a04 Chanlett, E.T. ENVIRONMENTAL PROTECTION. New York: McGraw-Hill, 1973. 569 p.

Designed as a text for undergraduate courses that deal with "man and his environment," this book is directed toward the "whys" of wastewater management as well as the management of air resources, solid wastes, food, and energy supplies. The levels of protection are categorized as (1) health effects, (2) convenience and aesthetics, and (3) ecology and ecosystems. Wastewater management is discussed in general terms as one of the aspects involved in the protection of environmental quality. (For general review, see Appendix A.)

706   Council on Environmental Quality. ENVIRONMENTAL QUALITY, FOURTH ANNUAL REPORT. Washington, D.C.: September 1973. 499 p.

Directed toward the total environment, this comprehensive report devotes considerable space and detail to water pollution and wastewater management. An exciting case of environmental clean-up is documented for the Willamette River in Oregon. An overview of environmental programs is presented, and trends in environmental quality are examined. The role of the citizen is described in terms of past effects and future opportunities.

Ecology and Ecosystems

707   Ford, R.F., and Hazen, W.E., eds. READINGS IN AQUATIC ECOL-OGY. Philadelphia: W.B. Saunders Co., 1972. 397 p. Paperbound.

A diversified selection of papers on aquatic ecology and eco-systems is contained in this book. The tone is that of high science; however, each paper is prefaced with a concise abstract. The last two sections are of particular interest to readers concerned with wastewater management, dealing with nutrient cycles and associated problems of eutrophication. The first four sections contain twenty-two papers on "pure" biological aspects of population ecology, physiological and behavioral ecology, and community ecology. Biologists will find this book of great value.

708   Hynes, H.B.N. THE ECOLOGY OF RUNNING WATERS. Toronto: University of Toronto Press, 1970. 555 p.

All aspects of the environment of fresh-water biota and fishes are explored and, where possible, quantified. Physical and chemical characteristics of flowing water are defined in relation to the ecology of streams. This excellent and comprehensive book on the limnology of streams complements the author's earlier work on the pollution of streams. Extensive documentation and illustrations are presented for attached algae, plankton, and benthic organisms. Fish and other large vertebrates are also well documented, and an index of organisms is included.

Impact Assessments

709   Detwyler, T.R.   MAN'S IMPACT ON ENVIRONMENT.   New York:
      McGraw-Hill, 1971.   731 p.

      The author has selected fifty-two papers that are generally spe-
      cific in scope and content but quite readable, to assess man's
      current impact on elements of the total environment.   Although
      grouped under ten topical headings, none of the articles is on
      the same subject.   The eight papers dealing with man's water
      range from hydrologic effects of urban land use to oil pollution
      of the oceans.   Articles are timely as well as pertinent.   The
      spread of pesticides, microorganisms, and air pollutants on a
      global scale is addressed.   The reader will be stimulated by
      this variety, as well as by the case-study approach where spe-
      cific facts are presented.

710   Dickert, T.G., and Domeny, K.R., eds.   ENVIRONMENTAL IMPACT
      ASSESSMENT: GUIDELINES AND COMMENTARY.   Berkeley: Univer-
      sity of California Extension, 1974.   238 p.

      This book resulted from a September 1972 Conference on Im-
      proving the Environmental Impact Assessment Process.   In addi-
      tion to several noteworthy articles on past and present impact
      assessments, a compendium of federal and state guidelines and
      legislation is included.   Pertinent documents on judicial review,
      social impact, and economic-equity assessments are listed in the
      bibliography.   Early examples of impact assessments are ana-
      lyzed, and needs and solutions are expounded.

Section 8

LEGISLATION AFFECTING WASTEWATER MANAGEMENT

Section 8

LEGISLATION AFFECTING WASTEWATER MANAGEMENT

State and federal legislation is the framework on which the water pollution control effort is based. In this section, references are cited in which the principal state and federal laws and regulations affecting wastewater management are summarized. Sources which contain important court cases establishing precedents that affect water pollution control are reviewed.

The current national water pollution control program is based primarily on the Federal Water Pollution Control Act Amendments of 1972. Under this far-reaching legislation, a timetable for providing specific levels of wastewater treatment on a nationwide basis has been established. Funds in the form of construction grants have been allocated as part of the law to aid in meeting the goals. Review of program progress to date indicates that the cost of achieving the goals established by the law may not be within the country's means at this time. The implementation timetable may be revised, depending on the results of current studies to assess the cost of implementation.

Although it is everyone's responsibility to minimize or prevent water pollution, it is government's responsibility to effect and enforce a program of water pollution control. Current programs and regulations are the result of both state and federal legislation.

A. LEGAL CONTROL

Legal control for water pollution control rests with the various state and federal agencies established by legislation dealing with the control of environmental quality. Interpretation of the various laws is often ultimately based on court rulings.

Documentation

*   Code of federal regulations
*   State water pollution control agencies

113

Unresolved Issues

* Effluent versus stream standards
* Suits involving wastewater management

801 Baldwin, R.B., ed. UCD LAW REVIEW, VOLUME 1: LEGAL CONTROL OF WATER POLLUTION. Davis: School of Law, University of California at Davis, 1969. 273 p.

> Written by law students, the various levels and forms of legal control of water pollution are identified and discussed in this review. State control of water pollution is discussed, with reference to California as a model. Control through interstate agreement, federal programs, and the private sector are subsequently discussed. Separate chapters on groundwater pollution and ocean pollution are included. The review is not current with respect to recent federal programs, particularly the 1972 Water Pollution Control Act Amendments, but in-depth coverage of the many facets of legal control of water pollution is provided.

802 Ross, S.S., ed. ENVIRONMENTAL REGULATION HANDBOOK. New York: Environmental Information Center, 1973.

> This handbook is designed to be a working reference. Federal and state regulations are compiled and organized for quick retrieval. It is divided into four sections: (1) the introductory section in which a guide to government agencies and a directory of pollution control officials are given, (2) the regulatory section, in which federal and state laws and regulations concerned with the environment are compiled, (3) the index, and (4) the retrieval section. All aspects of environmental control are covered, including water pollution. The handbook is updated continuously as changes occur. A valuable and complete reference.

B. PRINCIPAL PUBLIC LAWS

Surprisingly, water pollution control in the United States is derived from a limited number of public laws. Interpretations of these laws fill endless volumes.

Documentation

* National Environmental Protection Act (NEPA)
* Public Law 92-500
* Individual state acts and laws and local rules and ordinances

Unresolved Issues

* Interpretation of best practicable treatment technology and other provisions
* Court tests of various federal and state laws

803 Committee on Public Works, U.S. House of Representatives. LAWS OF THE UNITED STATES RELATING TO WATER POLLUTION CONTROL AND ENVIRONMENTAL QUALITY. Washington, D.C.: Government Printing Office, March 1973. 522 p. Paperbound.

A compilation of legislation developed by the Committee on Public Works in the field of water pollution control and water quality. The Federal Water Pollution Control Act Amendments of 1972 (Public Law 92-500) are presented in their entirety, with useful marginal notations to identify subject areas of the text.

804 National Water Commission. A SUMMARY-DIGEST OF THE FEDERAL WATER LAWS AND PROGRAMS. Washington, D.C.: Government Printing Office, May 1973. 205 p. Paperbound.

Federal water laws and programs are summarized concisely in this digest. The programs and statutory responsibilities of all federal agencies concerned with water resources are described. Thus water pollution control laws and programs represent only a part of those described. The digest was written primarily for the use of nonlawyer legislators and administrators as well as lawyers, but it would be a valuable reference for anyone seeking information regarding federal agency responsibilities.

805 National Water Commission. A SUMMARY-DIGEST OF STATE WATER LAWS. Washington, D.C.: Government Printing Office, May 1973. 826 p. Paperbound.

A companion document to the SUMMARY-DIGEST OF FEDERAL WATER LAWS AND PROGRAMS. This two-part digest presents and clarifies general and specific information on the water laws of each state. The development of state water laws and the organizational structure of state water law agencies are traced and discussed in the first part. Separate chapters dealing with surface and groundwater laws, interstate water rights, and federal and Indian-reserved water rights are also included in part I. The second part contains fifty chapters in which the water laws of each state are described. This document is a valuable reference, particularly on water rights. Information on water pollution or quality control laws is a small part of what is presented.

Appendix A

TEXTBOOKS AND REFERENCE WORKS

Appendix A

TEXTBOOKS AND REFERENCE WORKS

The following textbooks and reference works have been selected as representative of the best of those available dealing with the broad subject of wastewater management. Because of the rapid development and many changes now occurring in this field, much of the material contained in the various texts and reference books will, of necessity, be dated. To minimize this occurrence, an effort has been made to select works dealing with basic concepts and principles rather than those dealing with descriptions of specific treatment plants and facilities.

a01   Babbitt, H.E., and Baumann, R.E.   SEWERAGE AND SEWAGE TREATMENT.   8th ed.   New York: John Wiley & Sons, 1958.   797 p.

> The first version of this well-known text appeared in 1922. Until the early sixties, this text in its various editions was used extensively in undergraduate courses dealing with sewerage and sewage treatment. Unfortunately, much of the material in the eighth edition is now dated. Even so, much useful data and information may be found in this text. The first thirteen chapters deal with the estimation of the quantity of sewage and the design, construction, and maintenance of sewers. The remaining sixteen chapters are devoted primarily to the treatment of wastewater. A chapter on plant operation and maintenance is also included. (Also cited in Section 2, A and Section 3, C.)

a02   Bolton, R.L., and Klein, L.   SEWAGE TREATMENT: BASIC PRINCIPLES AND TRENDS.   2nd ed.   Ann Arbor, Mich.:   Ann Arbor Science Publishers, 1971.   260 p.

> This is an English book that has been reprinted in this country. Devoted primarily to the treatment and disposal of sewage, the level of coverage is elementary. It is heavy on descriptive material and rather light on quantitative applications. Easy to read, it can be used effectively as a primer for someone interested in knowing more about the treatment of sewage.

a03   Brock, T.D.   PRINCIPLES OF MICROBIAL ECOLOGY.   Englewood Cliffs, N.J.:   Prentice-Hall, 1966.   320 p.

The basics of microbial ecology are presented and discussed in a clear and readable manner in this beginning text. The reason that it is included in this listing is that it contains much useful information from which a clearer understanding of the microbial interactions and responses that occur in the biological processes used for wastewater treatment can be gained. The sections that deal with the development and growth of filamentous microorganisms will be of special interest to sanitary engineers involved in the design of activated sludge treatment systems. This text should be on the recommended reading list for all sanitary engineers.

a04   Chanlett, E.T. ENVIRONMENTAL PROTECTION. New York: McGraw-Hill, 1973. 569 p.

A general introductory text dealing with the subject of environmental protection and encompassing many of the same subjects discussed by Ehlers and Steel in MUNICIPAL AND RURAL SANITATION (a08). In addition to dealing with standard topics such as environmental quality, epidemiology, water resources, wastewater disposal, the air environment, and solid wastes management, it devotes separate chapters to such subjects as vector control, food protection, ionizing radiation, electromagnetic energy, and the energies of heat and sound. This would be a useful text for a course on the subject of man and his environment. It would also be a good general introduction to the study of sanitary science. (Also cited in Section 7, B.)

a05   Clark, J.W.; Viessman, W., Jr.; and Hammer, M.J. WATER SUPPLY AND POLLUTION CONTROL. 2nd ed. Scranton, Pa.: International Textbook Co., 1971. 674 p.

Both water supply and pollution control are considered in this volume, which was designed for use as an undergraduate text. Because it tries to cover far too much material in one volume, the treatment given some topics in individual sections tends to be spotty. Also, because of the joint coverage, the continuity tends to suffer. This is nevertheless a useful text, and it has achieved a reasonable measure of success. The inclusion of problems at the end of chapters increases its utility.

a06   Culp, R.L., and Culp, G.L. ADVANCED WASTEWATER TREATMENT. New York: Van Nostrand Reinhold Co., 1971. 310 p.

This is one of the first reference books in which the subject of advanced wastewater treatment is considered separately and in detail. Much of the material is drawn from the senior author's experience as general manager of the South Tahoe Public Utility Districts during the periods of development, start-up, and operation of the experimental wastewater reclamation plant located at South Lake Tahoe. For the most part, the treatment processes considered are similar to those used in industrial treatment and processing applications and the treatment of public water supplies. They include chemical coagulation and flocculation, sedimentation, stripping, recarbonation, filtration, activated carbon adsorption, carbon regeneration, solids handling, and a review of other possible methods. A useful chapter on selecting and combining unit processes to obtain any desired water quality is also included. Although not design oriented, this reference work is useful and informative on the general subject of advanced wastewater treatment. (Also cited in Section 3, F.)

a07 Eckenfelder, W.W., Jr. WATER QUALITY ENGINEERING FOR PRAC-
TICING ENGINEERS. New York: Barnes & Noble Books, 1970. 340
p. Paperbound.

This paperback contains much of the material used by the author in various short courses he has taught over the past ten years on the subject of water quality control. Material is presented on (1) the basic principles and theories of water pollution control, (2) the application of various unit operations and processes to specific treatment problems, (3) the significant parameters in water quality engineering, and (4) design procedures for many of the commonly used wastewater treatment operations and processes. Although many equations are presented, their application is not always delineated or well documented. The same is true for much of the data and information contained in the numerous summary tables included in this book. The book is of most value to someone with a working knowledge of the field.

a08 Ehlers, V.M., and Steel, E.W. MUNICIPAL AND RURAL SANITA-
TION. 6th ed. New York: McGraw-Hill, 1965. 677 p.

One of the few texts available in which the subject of municipal and rural sanitation is discussed. Although quite general, this text can be used to gain an overview of the problems of municipal and rural sanitation and some time-honored solutions. It should be followed by specific readings in areas of interest.

a09 Environmental Protection Agency. GUIDANCE FOR FACILITIES PLAN-
NING. 2nd ed. Washington, D.C.: 1974. 115 p. Paperbound.

All you ever wanted to know about preparing a facilities planning report in conformance with EPA regulations. Facilities plans are required of those who seek grants for construction of publicly owned waste treatment works under Title II of the Federal Water Pollution Control Act (Public Law 92-500). The general techniques of facilities planning and a suggested structure for such plans are outlined in this guide. This report will be revised periodically to incorporate significant policy changes and to keep it generally updated. (Also cited in Section 1, C and D and Section 6, C.)

a10   Fair, G.M., and Geyer, J.C. WATER SUPPLY AND WASTEWATER
       DISPOSAL. New York: John Wiley & Sons, 1954. 985 p.

       A valuable reference if you already know all about sanitary engi-
       neering. Although this reference was used for many years as a
       teaching text, it is not especially well suited for the purpose,
       principally because it is difficult to follow and contains numerous
       typographical errors. The fact that both water and wastewater are
       considered in this reference also adds to the confusion. It neverthe-
       less remains a milestone in the field of sanitary engineering. It re-
       presented the bringing together and rigorous, scientific presentation
       of a considerable body of knowledge dealing with sanitary engineer-
       ing. In this sense, it can be considered the modern counterpart of
       the original three-volume work, AMERICAN SEWERAGE PRACTICE,
       by Metcalf and Eddy (a18, a19). In terms of organization, the first
       half of the book, which includes thirty chapters, is devoted to such
       subjects as the statistical analysis of quantitative information, sur-
       face and ground water hydrology, collection of surface and ground
       water, and distribution of water. The latter half is devoted prin-
       cipally to the unit operations and processes used for water and
       wastewater treatment. Although it is difficult to extract much of
       the information presented for practical application, this reference
       is still recommended to the serious student of sanitary engineering.
       The revised version of this reference is also reviewed (a11).

a11   Fair, G.M.; Geyer, J.C.; and Okun, D.A. WATER AND WASTEWA-
       TER ENGINEERING. 2 vols. New York: John Wiley & Sons, 1968.
       Vol. 1, 650 p.; vol. 2, 668 p.

       This two-volume work is based on the book, WATER SUPPLY AND
       WASTEWATER DISPOSAL, written by Fair and Geyer and first
       published in 1954 (a10). Material related to the development of
       water supplies and to the design of wastewater collection systems is
       delineated in the first eighteen chapters that comprise the first vol-
       ume, which is subtitled "Water Supply and Wastewater Removal."
       Optimization techniques and a general discussion of engineering
       projects are also included in the first volume. Material related to
       the physical, chemical, and biological management of water re-
       sources is considered in the twenty chapters that comprise volume
       2, subtitled "Water Purification and Wastewater Treatment and
       Disposal." While this two-volume edition is, in many respects, an
       improvement over its predecessor, it remains difficult to use for
       several reasons: the style is somewhat pedantic, water supply and
       treatment are considered together with waste water collection
       treatment and disposal, the illustrative examples are difficult to
       follow, and the text material is not well related to practice.
       Nevertheless, these two volumes contain much useful information
       that can be extracted if the reader is already knowledgeable in
       the field. (Also cited in Section 3, A, B, E, and H.)

a12   Gloyna, E.F., and Eckenfelder, W.W., Jr., eds. WATER QUALTIY

IMPROVEMENT BY PHYSICAL AND CHEMICAL PROCESSES. Austin: University of Texas Press, 1970. 448 p. (Published for the Center for Research in Water Resources.)

The third in a series of twelve projected volumes to be published as a summary of a continuing series of special lectures and seminars, the present volume is devoted to the physical and chemical processes that are used to achieve and maintain water quality. The material has been organized into sections dealing with (1) water quality requirements for reuse, (2) settleable and suspended solids removal, (3) chemical treatment, and (4) sludge handling and disposal. This can be a valuable reference, particularly for design engineers. (Also cited in Section 3, E; Section 4, A; and Section 5, A.)

a13 Gurnham, C.F., ed. INDUSTRIAL WASTEWATER CONTROL. New York: Academic Press, 1965. 486 p.

Because industrial wastes are often found in domestic wastewater, it is important to have an understanding of their generation and nature. For this reason, this volume and the one by Nemerow (a23) have been included in this reference list. The twenty-five individual contributions that comprise this volume are organized in six sections dealing with (1) animal food products, (2) vegetable food products, (3) mining, (4) industries. In general, each industry or industry group is described, the wastes are identified, and the treatment methods in common use are reviewed. Although the material presented in this volume is largely descriptive, some useful tables are included. Furthermore, while the quality and usefulness of the individual sections is variable, the material in this volume is still useful as an introduction to the subject.

a14 Imhoff, K., and Fair, G.M. SEWAGE TREATMENT. 2nd ed. New York: John Wiley & Sons, 1956. 345 p.

The second edition of this book, first published in 1940, was completely rewritten because of the many changes and developments that occurred since then. Intended to reach a broad spectrum of readers, the book outlines the general nature of the facilities and processes used in sewage treatment as simply as possible. Unfortunately, the various subjects dealing with sewage treatment are presented in a discrete fashion and are not well related chapter to chapter. Also, much has occurred in the field since the publication of the second edition.

a15 Imhoff, K.; Muller, W.J.; and Thistlethwayte, D.K.B. DISPOSAL OF SEWAGE AND OTHER WATERBORNE WASTES. 2nd ed. London: Butterworth & Co., 1971. 415 p.

The origin of this text can be traced to 1906 when the senior author (now deceased) wrote the original German version. The "first edition" of this reference that appeared in 1956 was actually a translation of the sixteenth edition of the original German version. The material is organized into four major sections dealing with (1) introductory material and basic data, (2) methods of disposal, (3) the treatment of sewage and other waterborne wastes, and (4) the disposal of wastes from individual residences, institutions, and small communities. The various operations and processes are described in a direct manner, and the accompanying computations are straightforward. This book contains a great deal of useful information and is recommended as a practical reference.

a16 McGauhey, P.H. ENGINEERING MANAGEMENT OF WATER QUALITY. New York: McGraw-Hill, 1968. 302 p.

The material contained in this book was used originally as lecture notes for the first in a series of graduate courses dealing with sanitary and water resources engineering. In bringing it together in the one volume, the intent was to stimulate thought on the significance of water quality as a concept, and on the application of water quality concepts to water quality and water resources management. In keeping with the original intent, a wide variety of topics dealing with water quality is presented and discussed. The chapter dealing with engineered soil systems for water quality management is especially interesting. Because of its presentation format, unfortunately, a number of loose ends exist, and it is difficult for one inexperienced in the field to effectively associate all the concepts presented. Nevertheless, this text is recommended for review where a basic understanding of water quality concepts is necessary. (Also cited in Section 4, B and C.)

a17 McKinney, R.E. MICROBIOLOGY FOR SANITARY ENGINEERS. New York: McGraw-Hill, 1962. 293 p.

This book was written to teach sanitary engineering students the fundamentals of microbiology; it has been used extensively for the purpose. Now somewhat dated, it is no longer used as often and has been largely replaced with standard texts in microbiology written by microbiologists. The material presented in this text is divided into two principal parts: the first part is devoted to fundamental microbiology, the second to applied microbiology. The material in the first part is similar to what would be found in any standard text on microbiology, whereas the material in the second part is related primarily to biological waste treatment. Separate chapters are devoted to trickling filters, activated sludge, oxidation ponds, anaerobic digestion, and reuse disposal. The subjects of radioactivity, industrial wastes, milk and food, and air microbiology are also addressed in the second part. (Also cited in Section 3, B.)

a18 Metcalf, L., and Eddy, H.P. AMERICAN SEWERAGE PRACTICE,

VOL. 1: DESIGN OF SEWERS. 2nd ed. New York: McGraw-Hill, 1928. 774 p.

The publishing of AMERICAN SEWERAGE PRACTICE in three volumes between 1914 and 1915 represents a major milestone in the development of the sanitary engineering field. For the first time, material avaliable from various scattered sources, material from the files of Metcalf & Eddy Engineers, and material developed specifically for the purpose was brought together in one source. This three-volume issue was an immediate success and the popularity of volumes 1 and 3, which dealt with the design of sewers and the disposal of sewage, continued through several editions until the early 1940s, when both volumes went out of print. Although dated, much of the material contained in volume 1 dealing with the design of sewers is still usable. Where available, the entire three-volume series contains much material of historical value. It is also interesting to thumb through the pages of these old reference works to try and determine which of the many operations and processes that are described will be rediscovered next.

a19   . AMERICAN SEWERAGE PRACTICE, VOL. 3: DISPOSAL OF SEWAGE. 3rd ed. New York: McGraw-Hill, 1935. 909 p.

Of the three volumes comprising the work, AMERICAN SEWERAGE PRACTICE, volume 3, which dealt with the disposal of sewage, was perhaps the most popular. As evidenced by the publication of a third edition of the original text, this aspect of wastewater management underwent great change in the period between 1915 and 1935. Many unit operations and processes were tried during the period. Some were adopted, but most were left to the pages of history. Both the successes and failures of the period are documented in this reference work.

a20   . SEWERAGE AND SEWAGE DISPOSAL. 2nd ed. New York: McGraw-Hill, 1930. 799 p.

After the three volumes of AMERICAN SEWERAGE PRACTICE were published (a18, a19), the authors were urged to prepare a combined single volume for use as a classroom text. A condensed single volume was first issued in 1922. Unfortunately, the first edition of this text was not especially good. It suffered from being too general. Recognizing that something had to be done, R.G. Tyler from MIT and G.M. Fair from Harvard were enlisted; and the first edition was completely rewritten. The second edition remained the most widely adopted classroom text for the next ten to fifteen years, until it too went out of print. In terms of organization, the first ten chapters were devoted principally to the estimation of the quantity of sewage and to the design and construction of sanitary and storm sewers. The remaining eleven chapters dealt with the treatment and disposal of wastewater and sludge. Chapter 13, which dealt with sewage disposal by irrigation, is of special interest today because of the current emphasis on land disposal.

a21   Metcalf & Eddy, Inc.   WASTEWATER ENGINEERING: COLLECTION,
TREATMENT, DISPOSAL.   San Francisco: McGraw-Hill, 1972.   795 p.

After the publication of the second edition of SEWERAGE AND
SEWAGE DISPOSAL (a20), the firm of Metcalf & Eddy, Inc.
was asked repeatedly to issue a revised edition. After review-
ing the matter carefully, it was decided that the old text could
not be revised, and that an entirely new one would be needed
to encompass the numerous developments that have occurred in
the field in the last four decades. Since its publication, this
new text has been widely accepted and is now perhaps the most
well known in the field. Designed primarily as a teaching
text, it has also been widely adopted by practicing engineers.
The material is arranged in a logical sequence dealing with
collection, treatment, and disposal. Separate chapters on pumps
and pump stations, advanced wastewater treatment, and the conduct of wastewater treatment studies are also included in the
sixteen chapters that comprise this book. From an educational
standpoint, the material is presented so that the student will be
able to apply the fundamentals to the solution of practical problems. If a single text dealing with wastewater management is
to be purchased, this would be the choice. (Also cited in Section 2, A; Section 3, A, B, C, E, F, and H; Section 4, B;
and Section 5, A.)

a22   Mitchell, R., ed.   WATER POLLUTION MICROBIOLOGY.   New York:
Wiley-Interscience, 1972.   426 p.

This volume comprises seventeen individual papers from twenty-
three authors. The individual contributions are organized in
six sections dealing with (1) microbial changes induced by inorganic pollutants, (2) microbial changes induced by organic
pollutants, (3) intestinal pathogens as pollutants, (4) pollution
and community ecology, (5) microbial parameters of pollution,
and (6) microbiological approaches to pollution control. As is
often the case with books made up of individual contributions,
the treatment and coverage is spotty. With a price tag of
$19.50 in 1974, this volume may remain principally a library
reference.

a23   Nemerow, N.L.   LIQUID WASTE OF INDUSTRY: THEORIES, PRAC-
TICES, AND TREATMENT.   Reading, Mass.:   Addison-Wesley Publishing
Co., 1971.   584 p.

As previously noted, this and the reference by Gurnham (a13)
were included because industrial wastes are so often a component of domestic wastewater. The twenty-seven chapters that
comprise this volume have been organized into four major sections dealing with (1) basic knowledge and practices, (2) basic
theories of waste treatment, (3) application of basic knowledge

and theories, and (4) a review of major industrial wastes. The material included in the first three sections that occupy about half of the book is more or less standard fare and can be found in a number of texts and references. The second half, in which the major industrial wastes are considered, is of greater value, especially when some general background information is needed on a given industry. Although numerous references are presented for each industry, their inclusion is of questionable value because the reader has no basis for making a judgment concerning the usefulness of any individual reference and usually does not have the time or inclination to review all of the references. Also because each industrial waste situation is different, each must be considered separately. (Also cited in Section 3, A.)

a24   Phelps, E.B. STREAM SANITATION. New York: John Wiley & Sons, 1944. 287 p.

A gem--one of the nicest little books available in the entire field of sanitary engineering, a must for every sanitary engineer's bookshelf. Although it is out of print, look for it in used bookstores. This is the kind of book one can write when one clearly understands the subject with which he is dealing. In the case of Phelps, who was responsible for the original development of the oxygen sag analysis for streams, the subject of stream sanitation was second nature. A separate chapter on stream microbiology by J.B. Lackey is also noteworthy. The presentation dealing with the biochemical oxygen demand (BOD) should be read and reread by every sanitary engineer. (Also cited in Section 4, B.)

a25   Rich, L.G. UNIT OPERATIONS OF SANITARY ENGINEERING. New York: John Wiley & Sons, 1961. 308 p.

This and a companion volume on unit processing (a26) were the first books in the sanitary field in which the traditional approach of dealing with treatment methods and means in a process sequence was abandoned in favor of the unit approach, which is used commonly in the chemical engineering field. Although all of the important unit operations used for water and wastewater treatment were considered, the book never achieved wide popularity because (1) it was ahead of its time, (2) there was much confusion with respect to the usage of symbols and units, (3) many of the examples were difficult to follow, and (4) the presentation tended to be somewhat theoretical and was not well related to practice.

a26   _____. UNIT PROCESSES OF SANITARY ENGINEERING. New York: John Wiley & Sons, 1963. 199 p.

This is the companion volume to the book on unit operations (a25) by the same author. The material in this small volume is divided into two parts dealing with (1) biological processes and (2) chemical and related physical processes. This was one of the first volumes in which subjects such as electrodialysis, dry combustion, and wet combustion were considered in more than a cursory manner. Review comments given for the volume on unit operations also apply to this work.

a27 _____. ENVIRONMENTAL SYSTEMS ENGINEERING. New York: McGraw-Hill, 1973. 502 p.

Designed for junior- or senior-level engineering students, this text stresses the systems approach in the analysis and solution of problems related to the control of the environment. In keeping with the systems approach, the mathematics of systems analysis and the use of computers is emphasized. Although the water environment is stressed, elements of air pollution and its control, solid waste management, and radiological health are also introduced. The individual sections are brief, but this book is useful both as a text and reference because of the number of subjects considered and the rigorous approach used in their analysis. Perhaps the book's major shortcoming is that the material is not related closely to practice. Classroom lectures should bridge this gap in most situations.

a28 Salvato, J.A., Jr. ENVIRONMENTAL ENGINEERING AND SANITATION. 2nd ed. New York: Wiley-Interscience, 1972. 936 p.

Along with Ehlers and Steel (a08), this is one of the few texts dealing with the general subject of sanitation. It also includes separate chapters on environmental engineering planning, water supply, wastewater treatment and disposal, solid waste management, and air pollution control. A key feature of this text is the coverage given to small systems, including individual on-site disposal facilities. An excellent reference on the subject, especially for small communities, written by an engineer with extensive experience in this field.

a29 Sawyer, C.N., and McCarty, P.L. CHEMISTRY FOR SANITARY ENGINEERS. 2nd ed. San Francisco: McGraw-Hill, 1967. 518 p.

The 1960 publication of the first edition of this text under Sawyer's sole authorship was long accepted as the standard in the field. This revised edition is even more popular. The material is organized into two parts: the fundamental concepts of chemistry are presented and illustrated in the first; analysis of the specific constituents of importance found in water and wastewater is considered in the second. The sanitary significance of the various constituents in water and wastewater is also discussed in part 2. This book should be on every sanitary engineer's bookshelf. It also can be used as an introduction to the more advanced work, AQUATIC CHEMISTRY by Stumm and Morgan (a30). (Also cited in Section 3, B, H, and J.)

a30 Stumm, W., and Morgan, J.J. AQUATIC CHEMISTRY. New York:
Wiley-Interscience, 1970. 598 p.

The most up-to-date and complete reference text available in
one volume on the general subject of aquatic chemistry. To
utilize this text effectively, one should have a reasonably good
background in general chemistry. The text by Sawyer and
McCarty (a29) or a good general text in freshman chemistry can
be used as an introduction to this volume. In addition to .
standard fare dealing with chemical thermodynamics, acids and
bases, dissolved carbon dioxide, precipitation and dissolution,
and oxidation and reduction, separate chapters are devoted to
such interesting subjects as metal ions in aqueous solution, the
regulation of the chemical composition of natural waters, the
solid-solution interface, and case studies dealing with phosphorus, iron, and manganese. An excellent reference for the advanced student and worker involved with problems in aquatic
chemistry. Unfortunately, the cost of this book ($24.95 in
1974) has limited its wide acceptance. Although a paperback
version is available, it tends to fall apart with continued usage.

a31 Velz, C.J. APPLIED STREAM SANITATION. New York: Wiley-Interscience, 1970. 619 p.

This text is intended as a complementary contribution to
STREAM SANITATION by Phelps (a24). Essentially, most of
the topics covered are also contained in STREAM SANITATION,
but they are treated more extensively in this volume. In addition to material dealing specifically with stream sanitation,
separate chapters are devoted to self-purification in estuaries,
the impact of river development on waste assimilation, the efficient use of assimilation capacity, the conduct of stream surveys, and the management of waste disposal and stream flow
quantity and quality. This is a useful reference if you deal
specifically with stream sanitation; otherwise it tends to be a
bit ponderous. For those whose specialty is not stream sanitation,
this book will still be of interest because of the inclusion of two
appendices that deal with statistical analysis and bacterial
enumeration. In the appendix entitled "Statistical Tools," the
graphic approach to statistics is presented and illustrated. This
is one of the few books in which this information may be found.
Similarly, the basic mathematical theory of the dilution method for
estimating bacterial numbers is delineated in the appendix entitled "Bacterial Enumeration."

a32 Weber, W.J., Jr. PHYSICOCHEMICAL PROCESSES FOR WATER QUAL-
ITY CONTROL. New York: John Wiley & Sons, 1972. 666 p.

As indicated in the title, the coverage in this volume is limited
to physical and chemical processes that are used commonly for
water quality control. The subjects considered include process
dynamics; reactions and reactors; coagulation and flocculation;

sedimentation; filtration; adsorption; ion exchange; membrane processes; chemical oxidation; disinfection; corrosion and corrosion control; aeration and gas transfer; and sludge treatment. Since it is the combined effort of eight collaborators, the writing tends to be somewhat uneven as does the coverage given to some of the individual topics. The level of the presentation is appropriate for advanced undergraduates and graduate students. Because the material presented is not often related to practical examples, this text is best suited for use in a more theoretical course. (Also cited in Section 3, B, E, F, and H and Section 5, A.)

Appendix B

JOURNALS AND PERIODICALS

Appendix B
JOURNALS AND PERIODICALS

Over the past twenty years the number of journals and periodicals dealing with subject matter related to wastewater engineering has increased significantly. In the following listing, only those journals and periodicals that have proven to be of most value to workers in the field have been included. Some useful foreign journals and publications are also listed. If not readily available, they usually can be secured through interlibrary loans.

APPLIED MICROBIOLOGY - American Society for Microbiology, 1913 I Street N.W., Washington, D.C. 20006.

> Published monthly. This journal contains original papers and reviews dealing principally with the metabolism and physiology of microorganisms.

BIOTECHNOLOGY AND BIOENGINEERING - John Wiley & Sons, Inc., Twentieth and Northampton Streets, Easton, Pennsylvania 18042.

> Published bimonthly. This journal covers a wide range of subjects dealing with both applied and basic research. Papers and articles on technology and theory of waste disposal and microbial food production are included routinely.

EFFLUENT AND WATER TREATMENT JOURNAL - Thunderbird Enterprise, 3 Clements Inn, London, W.C. 2, England.

> Published monthly. This English journal aims its articles at the practicing engineer; workable solutions to pollution problems are stressed.

ENVIRONMENTAL POLLUTION - Elsevier Publishing Co., Ltd., Ripple Road, Barking, Essex, England.

> Published monthly. This international journal deals principally with the biological and ecological effects of all types of environmental pollution.

ENVIRONMENTAL SCIENCE AND TECHNOLOGY - American Chemical Society, 1155 Sixteenth Street N.W., Washington, D.C. 20036.

> Published monthly. The technical articles in this journal tend to be more theoretically oriented. Usually a feature article on some aspect of environmental pollution and control, written in layman's terms, is included.

INDUSTRIAL WASTES - Scranton Publishing Co., 454 South Wabash Avenue, Chicago, Illinois 60606.

> Published bimonthly. This journal emphasizes providing practical and technical information and data dealing with the solution of industrial waste disposal problems. Columns dealing with new products, people, literature, meetings, and books are included regularly.

JOURNAL OF ENVIRONMENTAL HEALTH - National Environmental Health Association, 1600 Pennsylvania, Denver, Colorado 80203.

> Published bimonthly. This journal gives rather broad coverage, including topics varying from noise pollution to animal disease. Because of its close relationship to environmental health, articles on wastewater management are also included.

JOURNAL OF ENVIRONMENTAL QUALITY - American Society of Agronomy, Crop Science Society of America, Soil Science Society of America, 677 South Segoe Road, Madison, Wisconsin 53711.

> Published quarterly. The articles in this journal are technically oriented and generally deal with environmental effects of agriculture. Topics range from pesticide and nutrient problems to land application of wastes.

JOURNAL OF FERMENTATION TECHNOLOGY - The Society of Fermentation Technology, Faculty of Engineering, Osaka University, Higashinoda, Miyakojima, Osaka, Japan.

> Published monthly. Most of the articles are written in Japanese, although the titles, abstracts, and graph and table titles in the various articles dealing with fermentation technology and waste treatment are usually in English.

JOURNAL OF THE SANITARY ENGINEERING DIVISION ASCE - American Society of Civil Engineers, 345 East Forty-seventh Street, New York, New York 10017.

> Published bimonthly. This journal and the JOURNAL OF THE WATER POLLUTION CONTROL FEDERATION are two of the principal sources of information and data for workers in the field of wastewater engineering. The articles cover a broad spectrum of subject matter related to sanitary engineering, are usually well documented, and are open to discussion by the membership at large.

JOURNAL OF THE WATER POLLUTION CONTROL FEDERATION - Water Pollution Control Federation, 3900 Wisconsin Avenue, Washington, D.C. 20016.

Published monthly. This journal and the JOURNAL OF THE SANITARY ENGINEERING DIVISION OF ASCE (above) are two of the principal sources of information for workers in the field of wastewater engineering. Prior to adopting the present title, this journal was published under the title of SEWAGE WORKS JOURNAL (until 1950) and SEWAGE AND INDUSTRIAL WASTES (until 1960). Subject matter deals principally with the collection, treatment, and disposal of domestic and industrial wastes. In addition to articles on technical subjects, operation and maintenance are also well covered. A literature survey compiled by noted authorities in various areas of wastewater management is an annual feature. This extensive, well organized survey is a valuable source of references. Author, subject, and geographical indexes covering the years 1928-48, 1949-58, 1959-63, and 1964-68 are also available.

MUNICIPAL ENGINEERING - 178-202 Great Portland Street, London, WIH 6NH, England.

Published weekly. This English paper includes general coverage of a variety of subjects dealing with municipal engineering such as public and environmental health, refuse collection and disposal, and wastewater collection, treatment, and disposal.

POLLUTION ENGINEERING - Technical Publishing Co., 35 Mason Street, Greenwich, Connecticut 06830.

Published bimonthly. This periodical deals with the control of environmental pollution. Subject areas covered are varied and have included air, solid waste, wastewater, and noise. Useful information on available literature is also included.

PUBLIC WORKS - Public Works Journal Corp., East Stroudsburg, Pennsylvania 18301.

Published monthly. This periodical is aimed at engineers and administrators responsible for the design, construction, operations, and administration of public utility systems. Articles dealing with various aspects of wastewater engineering are routine fare.

THE PUBLIC HEALTH ENGINEER - Institution of Public Health Engineers, 32 Eccleston Square, Westminster, London, SWIV 1PB, England.

Published six times per year. This English journal routinely contains articles dealing with various aspects of wastewater engineering.

WATER, AIR, AND SOIL POLLUTION - D. Reidel Publishing Co., Dordrecht, Holland.

Published quarterly in Holland. This interdisciplinary journal deals with the various physical and biological operations and processes on the environment.

WATER AND POLLUTION CONTROL - Southam Business Publication, Ltd., 1450 Don Mills Road, Don Mills, Ontario, Canada.

Published monthly. This Canadian periodical is practically oriented, and both technical and local news items are included. Design, construction, and operation and maintenance of wastewater treatment facilities are stressed in the articles and features.

WATER AND SEWAGE WORKS - Scranton Publications, 35 East Wacker Drive, Chicago, Illinois 60601.

Published monthly. This periodical is oriented towards engineers and operators involved in the design, construction, and operation of water treatment facilities and pollution control facilities for both domestic and industrial wastes. Columns dealing with new products, people, literature, meetings, and books are regular fare.

WATER AND WASTE TREATMENT - D.R. Publications, Ltd., 103 Brigstock Road, Thornton Heath, Surrey, England.

Published monthly. This English publication is a trade journal directed towards engineers and operators in the waste treatment field.

WATER AND WASTES ENGINEERING - Reuben H. Donnelley Publications, 466 Lexington Avenue, New York, New York 10017.

Published monthly. This publication is similar to WATER AND SEWAGE WORKS (above), but the emphasis is more toward the water resources engineering profession. Nevertheless, a number of articles related to wastewater engineering are usually included.

WATER POLLUTION CONTROL - Institute of Water Pollution Control, 49-55 Victoria Street, London, S.W. 1, England.

Published bimonthly. This journal is the English counterpart of the JOURNAL OF THE WATER POLLUTION CONTROL FEDERATION. In general, the subject areas covered are similar.

WATER RESEARCH - International Association on Water Pollution Research, Maxwell House, Fairview Park, Elmsford, New York 10523.

Published monthly. This journal deals principally with original research in the field of water quality management and has an international authorship and readership. Every aspect of water quality management, ranging from basic chemistry to advancements in waste treatment control systems, is included. This is a useful journal because it provides insight into international thought and developments.

Appendix C

NEWSLETTERS

Appendix C
NEWSLETTERS

As a result of the active role of the federal government in maintaining the quality of the environment, a number of newsletters are now being published. In general, these newsletters are designed to provide up-to-the-minute information on federal contracts and grants; new and proposed federal regulations; federal, state, and university research reports; people in the field; conferences and meetings; market trends; and equipment and new product development. Typically, the information contained in these newsletters is oriented mostly towards engineering managers and administrators. Sample copies can usually be obtained for review by writing to the circulation department of the various newsletters. To be as current as possible, most of the newsletters are published weekly. Yearly subscription rates vary from $8 to over $150.

AIR AND WATER NEWS - S.H. Brams, 801-7 New Center Building, Detroit, Michigan 48202.

> Published weekly. Local, national, and worldwide environmental news items are reviewed. Special emphasis is given to legislative actions. Information on people and products is also included.

AIR/WATER POLLUTION REPORT - Business Publishers, Inc., P.O. Box 1067, Blair Station, Silver Spring, Maryland 20910.

> News and information on air and water quality developments and legislation are standard fare in this weekly publication. Federal grants and a calendar of meetings are also included.

CLEAN AIR AND WATER NEWS - Commerce Clearing House, Inc., 4025 West Peterson Avenue, Chicago, Illinois 60646.

> Published weekly. News concerning the government's role in environmental legislation, research, policy, and related issues is the principal fare in this newsletter.

CLEAN WATER REPORT - Business Publishers, Inc., P.O. Box 1067, Blair Station, Silver Spring, Maryland 20910.

> Published since 1964, this biweekly is largely devoted to a review of progress on water quality legislation, research, contracts, and

grants. New products and conference listings are also included.

ENVIRONMENT REPORTER - The Bureau of National Affairs, Inc., 1231
Twenty-fifth Street N.W., Washington, D.C. 20037.

A weekly notification and reference service on current developments in pollution control and environmental protection; federal
laws, regulations, policies, and programs; state environmental
laws; and related issues.

ENVIRONMENTAL CONTROL NEWS FOR SOUTHERN INDUSTRY - Ramcon,
Inc., 223 Scott Street, Memphis, Tennessee 38112.

Published monthly. This newsletter deals with regulatory, legislative, and technical developments related to environmental quality control as well as the activities of environmentalist groups.

ENVIRONMENTAL ENGINEERING NEWSLETTER - 46 St. Clair Avenue East,
Toronto 290, Ontario, Canada.

Published monthly. News and information that are helpful in
making purchasing and marketing decisions in the pollution control
field are presented in summary form.

ENVIRONMENTAL LEGISLATION - 228 Parran Hall, Graduate School of
Public Health, University of Pittsburgh, Pittsburgh, Pennsylvania 15213.

Published weekly when Congress is in session. New regulations,
programs, and funding sources are reviewed and discussed. Action
on key bills and committee work is also documented.

ENVIRONMENTAL TECHNOLOGY AND ECONOMICS - Environmental Sciences Division, E.R.A., Inc., 750 Summer Street, Stamford, Connecticut
06901.

Published biweekly. Technical, economic, and legislative trends
and actions in air pollution control, solid waste, desalination, and
water renovation are presented in digest form. Federal, state,
and business news, along with information on meetings, government
work, and related issues may also be found in each issue.

FEDERAL REGISTER - U.S. Government Printing Office, Washington, D.C.
20402.

Published daily by the Government Printing Office, the FEDERAL
REGISTER is must reading for anyone interested in doing contract
and research work for the government. Because all new entries
in the Code of Federal Regulations are published here, the reader
needs a thorough knowledge of the category of regulations that
interest him. Using the index, he can locate the code numbers
of categories of interest.

NEW POLLUTION TECHNOLOGY - 6782 La Jolla Boulevard, Suite A, Box 191, La Jolla, California 92037.

Published biweekly. The publication's focus is on summary reports on research and developments in the pollution field. Information on contract awards and meetings is also included.

WASHINGTON ENVIRONMENTAL PROTECTION REPORT - Callahan Publications, P.O. Box 3751, Washington, D.C. 20007.

Published weekly. Financial considerations related to pollution control, especially government contracts, are the main fare of this newsletter.

WATER IN THE NEWS - Soap and Detergent Association, 475 Park Avenue South, New York, New York 10016.

Highlights of the latest developments in wastewater management, especially anything to do with detergents, are contained in this monthly newsletter.

WATER NEWSLETTER - Water Information Center, 44 Sintsink Drive, East Port Washington, Long Island, New York 19050.

Published biweekly. Water supply, waste disposal, and conservation are the principal topics considered in this newsletter.

Appendix D

ABSTRACTS AND DIGESTS

Appendix D

ABSTRACTS AND DIGESTS

Because the monthly output of literature in the field of wastewater engineering has expanded so greatly during the past twenty years, a number of organizations are now engaged in preparing abstracts and digests of this literature. Although these are useful in identifying the published literature, the problem of assessing the utility and applicability of the information contained in each article still remains. The abstracts and digests listed below are those that relate most to the field of wastewater engineering. Yearly subscription rates vary from $20 to over $200.

ASCATOPICS - Institute for Scientific Information, 325 Chestnut Street, Philadelphia, Pennsylvania 19106.

> Published weekly. Literature in the fields of medicine, biology, chemistry, and environmental science is reviewed.

ENVIRONMENT ABSTRACTS (formerly ENVIRONMENT INFORMATION ACCESS) - Environment Information Center, Inc., 124 East Thirty-ninth Street, New York, New York 10016.

> Published monthly. This information summary covers the broad field of environmental quality control. More than 4,500 technical, scientific, and general journals; major newspapers; and television and radio sources are reviewed in preparation for each issue. Document retrieval and research services are available to subscribers.

THE ENVIRONMENT INDEX - Environment Information Center, Inc., 124 East Thirty-ninth Street, New York, New York 10016.

> Published annually in December. This readers' guide provides an index to articles, books, films, legislation, environment-related patents, a directory of pollution control officials, and an overview of the year from the environmental standpoint.

ENVIRONMENTAL AWARENESS READING LIST - National Technical Information Service (NTIS), U.S. Department of Commerce, Springfield, Virginia 22151.

> Published semimonthly. Article titles and often a descriptive sen-

tence or two from more than one hundred publications dealing with environmental issues are given. The material is arranged alphabetically by author or title. A yearly index also includes a subject list.

ENVIRONMENTAL PERIODICALS - Environmental Studies Institute, International Academy at Santa Barbara, Rivera Campus, 2045 Alameda Padre Serra, Santa Barbara, California 93013.

Published irregularly. Each volume contains reproductions of the table of contents from more than 250 national and international journals dealing with environmental issues.

ENVIRONMENTAL POLLUTION AND CONTROL - National Technical Information Service (NTIS), U.S. Department of Commerce, Springfield, Virginia 22151.

Published monthly. Publications that are available for purchase from NTIS are announced in abstract form. The material offered is from more than 225 federal sources, including EPA, OSN, TVA, and the Federal Reserve Commission. It includes information on air, solid waste, noise, and water pollution.

I.C.E. ABSTRACTS - I.C.E. Abstracts, 26-34 Old Street, London, ECN 9AD, England.

Published ten times per year. Abstracts of articles appearing in more than ninety technical journals from all over the world are included. Articles are abstracted by qualified engineers in their native language and then translated into English. General subjects in civil engineering are covered.

POLLUTION ABSTRACTS - Oceanic Library and Information Center, P.O. Box 2369, La Jolla, California 92037.

Published bimonthly. Articles from both domestic and foreign technical reports, newspapers, journals, patents, symposia, and government documents are abstracted. The coverage is directed towards the effects of pollution on the environment.

SELECTED WATER RESOURCES ABSTRACTS - Water Resources Scientific Information Center, Office of Water Resources Research, U.S. Department of Interior Clearinghouse for Federal Scientific and Technical Information, Springfield, Virginia 22151.

Published bimonthly. Each abstract entry is classified into ten fields and sixty groups. Author, subject, organizational, and access number indexes are available.

WATER QUALITY CONTROL DIGEST - University Digest Service, P.O. Box 343, Troy, Michigan 48084.

Published bimonthly. Contains digests of articles and reviews related to water quality control from more than one hundred trade journals and magazines, government reports, and other indexes. The sources of information are worldwide.

WATER RESOURCES ABSTRACTS - American Water Resources Association, Urbana, Illinois 61801.

Published monthly. Articles are grouped in more than forty categories and abstracts. Although the publication is related more closely to water resources, articles on related topics in wastewater engineering are also included.

Appendix E

GOVERNMENTAL AGENCIES

Appendix E

GOVERNMENTAL AGENCIES

The principal governmental agencies responsible for water pollution control are the Environmental Protection Agency (EPA) of the federal government and the individual state agencies. The addresses of the EPA administrative headquarters, national environmental research centers, and regional offices are presented in Appendix E-1. The name and address of the responsible agency in each state is presented in Appendix E-2. Because of rapid changes in personnel assignments, the names of individuals are not given.

E-1   DIRECTORY OF ENVIRONMENTAL PROTECTION AGENCY

EPA Administrative Headquarters

Environmental Protection Agency
Waterside Mall Building
Fourth and M Streets S.W.
Washington, D.C.   20460

National Environmental Research Centers

P.O. Box 12055
Research Triangle Park, North Carolina   27711

200 Southwest Thirty-fifth Street
Corvallis, Oregon   97330

5555 Ridge Avenue
Cincinnati, Ohio   54268

P.O. Box 15027
Las Vegas, Nevada   89114

EPA Regional Offices and States Included in Each Region

Region 1

Room 2203
John F. Kennedy Federal Office Building
Boston, Massachusetts   02203
(Connecticut, Maine, Massachusetts, New Hampshire, Rhode Island, Vermont)

Region 2

Room 908
26 Federal Plaza
New York, New York 10007
(New Jersey, New York, Puerto Rico, Virgin Islands)

Region 3

Curtis Building
Sixth and Walnut Streets
Philadelphia, Pennsylvania 19106
(Delaware, Maryland, Pennsylvania, Virginia, West Virginia, District of Columbia)

Region 4

Suite 300
1421 Peachtree Street N.E.
Atlanta, Georgia 30309
(Alabama, Florida, Georgia, Kentucky, Mississippi, North Carolina, South Carolina, Tennessee)

Region 5

230 South Dearborn
Chicago, Illinois 60604
(Illinois, Indiana, Michigan, Minnesota, Ohio, Wisconsin)

Region 6

1600 Patterson Street
Suite 1100
Dallas, Texas 75202
(Arkansas, Louisiana, New Mexico, Texas, Oklahoma)

Region 7

1735 Baltimore Avenue
Kansas City, Missouri 64108
(Iowa, Kansas, Missouri, Nebraska)

Region 8

Room 900
Lincoln Tower Building
1860 Lincoln Street
Denver, Colorado 80203
(Colorado, Montana, North Dakota, South Dakota, Utah, Wyoming)

Region 9

100 California Street
San Francisco, California 94111
(Arizona, California, Hawaii, Nevada, American Samoa, Guam, Trust Territories of Pacific Islands, Wake Island)

Region 10

1200 Sixth Avenue
Seattle, Washington 98101
(Alaska, Idaho, Oregon, Washington)

E-2 STATE WATER POLLUTION CONTROL AGENCIES

ALABAMA

Alabama Water Improvement
 Commission
State Office Building
Montgomery, Alabama 36104

ALASKA

Alaska Department of Environmental
 Conservation
Pouch O
Juneau, Alaska 99801

ARIZONA

State Health Department
Division of Water Quality Control
Hayden Plaza West
4019 North Thirty-third Avenue
Phoenix, Arizona 85017

ARKANSAS

Department of Pollution Control
 and Ecology
8001 National Drive
Little Rock, Arkansas 72209

CALIFORNIA

State Water Resources Control Board
1416 Ninth Street
Sacramento, California 95814

COLORADO

Colorado Department of Health
Water Pollution Control Division
4210 East Eleventh Avenue
Denver, Colorado 80220

CONNECTICUT

Department of Environmental
 Protection
165 Capitol Avenue
Hartford, Connecticut 06115

DELAWARE

Department of Natural Resources and
 Environmental Control
Division of Environmental Control
Tatnall Building
Dover, Delaware 19901

DISTRICT OF COLUMBIA

Bureau of Air and Water Quality
 Control
Washington, D.C. 20002

FLORIDA

Department of Pollution Control
2562 Executive Center Circle
Montgomery Building
Tallahassee, Florida 32301

GEORGIA

Environmental Protection Division
Department of Natural Resources
270 Washington Street S.W.
Atlanta, Georgia 30334

HAWAII

Environmental Health Division
Department of Public Health
P.O. Box 3378
Honolulu, Hawaii 96801

IDAHO

Idaho Department of Health
Statehouse
Boise, Idaho 83707

ILLINOIS

Environmental Protection Agency
2200 Churchill Road
Springfield, Illinois 62706

INDIANA

Indiana Stream Pollution Control
Board
1330 West Michigan Street
Indianapolis, Indiana 46206

IOWA

Iowa Water Quality Commission
Department of Environmental Quality
Lucas State Office Building
Des Moines, Iowa 50319

KANSAS

Division of Environment
Department of Health and Environment
Building 740, Forbes Air Force Base
Topeka, Kansas 66620

KENTUCKY

Division of Water
Department for Natural Resources and
Environmental Protection
275 East Main Street
Frankfort, Kentucky 40601

LOUISIANA

Louisiana Stream Control Commission
P.O. Drawer FC-LSU
Baton Rouge, Louisiana 70803

MAINE

Environmental Improvement Commission
State House
Augusta, Maine 04330

MARYLAND

Department of Health and Mental
Hygiene
Environmental Health Administration
610 North Howard Street
Baltimore, Maryland 21201

MASSACHUSETTS

Division of Water Pollution Control
Leverett Saltonstall Building
100 Cambridge Street
Boston, Massachusetts 02202

MICHIGAN

Water Resources Commission
Stevens T. Mason Building
Lansing, Michigan 48926

MINNESOTA

Minnesota Pollution Control Agency
1935 West County Road B2
Roseville, Minnesota 55113

MISSISSIPPI

Mississippi Air & Water Pollution
Control Commission
P.O. Box 827
Jackson, Mississippi 39205

MISSOURI

Missouri Clean Water Commission
P.O. Box 154
Jefferson City, Missouri 65101

MONTANA

Department of Health and
Environmental Sciences
Cogswell Building
Helena, Montana 59601

NEBRASKA

Department of Environmental Control
1424 P Street
P.O. Box 94653
State House Station
Lincoln, Nebraska 68509

NEVADA

Environmental Protection Commission
201 South Fall Street
Carson City, Nevada 89701

NEW HAMPSHIRE

New Hampshire Water Supply and
Pollution Control Commission
105 Loudon Road
P.O. Box 95
Concord, New Hampshire 03301

NEW JERSEY

Division of Water Resources
New Jersey Department of
Environmental Protection
P.O. Box 1390
Trenton, New Jersey 08625

NEW MEXICO

Environmental Improvement Agency
Water Quality Division
P.O. Box 2348
Santa Fe, New Mexico 87503

NEW YORK

New York State Department of
Environmental Conservation
50 Wolf Road
Albany, New York 12201

NORTH CAROLINA

Water Quality Division Office of
Water and Air Resources
P.O. Box 27687
Raleigh, North Carolina 27611

NORTH DAKOTA

Division of Water Supply and
Pollution Control
North Dakota State Department of
Health
State Capitol Building
Bismarck, North Dakota 58501

OHIO

State of Ohio Environmental
Protection Agency
P.O. Box 1049
Columbus, Ohio 43216

OKLAHOMA

Oklahoma State Department of Health,
Water Quality Control Division
Northeast Tenth and Stonewall
Oklahoma City, Oklahoma 73105

OREGON

Department of Environmental Quality
1234 Morrison S.W.
Portland, Oregon 97205

PENNSYLVANIA

Bureau of Water Quality Management
Department of Environmental Resources
P.O. Box 2063
Harrisburg, Pennsylvania 17120

RHODE ISLAND

Division of Water Supply and Pollution
Control
Rhode Island Department of Health
209 Health Building
Davis Street
Providence, Rhode Island 02908

SOUTH CAROLINA

Office of Environmental Quality
Control
South Carolina Department of Health
and Environmental Control
J. Marion Sims Building
2600 Bull Street
Columbia, South Carolina 29201

SOUTH DAKOTA

Department of Environmental Protection
Office Building No. 2
Pierre, South Dakota 57501

TENNESSEE

Tennessee Water Quality Control Board
621 Cordell Hull Building
Nashville, Tennessee 37219

TEXAS

Texas Water Quality Board
P.O. Box 13246
Austin, Texas 78711

UTAH

Utah State Division of Health
44 Medical Drive
Salt Lake City, Utah 84113

VERMONT

Agency for Environmental Conservation
Department of Water Resources
Court Street
Montpelier, Vermont 05602

VIRGINIA

State Water Control Board
P.O. Box 11143
Richmond, Virginia 23230

VIRGIN ISLANDS

Department of Health
Charlotte Amalie
St. Thomas, Virgin Islands 00801

WASHINGTON

State of Washington
Department of Ecology
Olympia, Washington 98504

WEST VIRGINIA

Department of Natural Resources
Division of Water Resources
1201 Greenbrier Street
Charleston, West Virginia 25311

WISCONSIN

Bureau of Water Supply and
    Pollution Control
Box 450
Madisón, Wisconsin 53701

WYOMING

Wyoming Department of Health
State Office Building
Cheyenne, Wyoming 82002

Appendix F

PUBLISHERS AND PUBLISHERS' REPRESENTATIVES

Appendix F

PUBLISHERS AND PUBLISHERS' REPRESENTATIVES

Academic Press, Inc.
111 Fifth Avenue
New York, New York   10003

Addison-Wesley Publishing Co., Inc.
Jacob Way
Reading, Massachusetts   01867

Advanced Waste Treatment Research
   Laboratory
4676 Columbia Parkway
Cincinnati, Ohio   45206
(See also U.S. Environmental
   Protection Agency)

Air Pollution Control Association
4400 Fifth Avenue
Pittsburgh, Pennsylvania   15213

American Association for the
   Advancement of Science
1515 Massachusetts Avenue N.W.
Washington, D.C.   20005

American Association of Professors
   in Sanitary Engineering.   See
University of Illinois

American Chemical Society
1155 Sixteenth Street N.W.
Washington, D.C.   20036

American City Magazine
Berkshire Common
Pittsfield, Massachusetts   01201

American Concrete Institute
Publications Department
Box 19150
Detroit, Michigan   48219

American Concrete Pressure Pipe
   Association
1501 Wilson Boulevard
Arlington, Virginia   22209

American Conference of Governmental
   Industrial Hygienists
P.O. Box 1937
Cincinnati, Ohio   45201

American Dental Association
211 East Chicago Avenue
Chicago, Illinois   60611

American Elsevier Publishing Co., Inc.
52 Vanderbilt Avenue
New York, New York   10017

American Geophysical Union
1707 L Street N.W.
Washington, D.C.   20036

American Institute of Chemical
   Engineers
36 East Forty-seventh Street
New York, New York   10017

American Insurance Association
85 John Street
New York, New York   10038

Publishers

American Iron and Steel Institute
Engineering Division Library
150 East Forty-second Street
New York, New York   10017

American Petroleum Institute
1271 Avenue of the Americas
New York, New York   10020

American Public Health
  Association, Inc.
1015 Eighteenth Street N.W.
Washington, D.C.   20036

American Public Works Association
1313 East Sixtieth Street
Chicago, Illinois   60637

American Society for Metals
Metals Park, Ohio   44073

American Society for Testing and
  Materials
1916 Race Street
Philadelphia, Pennsylvania   19103

American Society of Agricultural
  Engineers
2950 Niles Road
St. Joseph, Michigan   49085

American Society of Agronomy
677 South Segoe Road
Madison, Wisconsin   53711

American Society of Civil Engineers
345 East Forty-seventh Street
New York, New York   10017

American Society of Mechanical
  Engineers
345 East Forty-seventh Street
New York, New York   10017

American Water Resources Association
Box 434
Urbana, Illinois   61801
Ohio Division: Neil L. Drobny,
  President
c/o Battelle Memorial Institute
Columbus, Ohio   43201

American Water Works Association
Publication Sales Department
2 Park Avenue
New York, New York   10016

Analytical Quality Control Laboratory
1014 Broadway
Cincinnati, Ohio   45202
(See also U.S. Environmental
  Protection Agency)

Ann Arbor Science Publishers, Inc.
P.O. Box 1425
Ann Arbor, Michigan   48106

Appalachian Regional Commission
1666 Connecticut Avenue
Washington, D.C.   20233

Appleton-Century-Crofts
292 Madison Avenue
New York, New York   10017

Arco Publishing Co.
219 Park Avenue, South
New York, New York   10003

Asian Institute of Technology
Henri Dunant Street
P.O. Box 2754
Bangkok, Thailand

Atheneum Press
(Halliday Lithograph Corp.)
Circuit Street
West Hanover, Massachusetts   02339

Baker Book House
1019 Wealthy Street S.E.
Grand Rapids, Michigan   49506

Ballantine Books
36 West Twentieth Street
New York, New York   10003

Barnes & Noble, Inc.
105 Fifth Avenue
New York, New York   10003

Bogden & Quigley, Inc.
19 North Broadway
Tarrytown-on-Hudson, New York
10591

Boston College Law School
Environmental Affairs, Inc.
Brighton, Massachusetts 02135

R.R. Bowker Co.
1180 Avenue of the Americas
New York, New York 10036

George Braziller, Inc.
1 Park Avenue
New York, New York 10016

Brigham Young University Press
205 University Press Building
Provo, Utah 84601

British Information Services
845 Third Avenue
New York, New York 10022

British Valve Manufacturers'
Association
14 Pall Mall
London, SWIY 51Z, England

William C. Brown Co., Publishers
2460 Kerper Boulevard
Dubuque, Iowa 52001

Bureau of National Affairs
1231 Twenty-fifth Street N.W.
Washington, D.C. 20037

Bureau of Solid Waste Management
Environmental Health Services
5555 Ridge Avenue
Cincinnati, Ohio 45213

Burgess Publishing Co.
426 Sixth Street
Minneapolis, Minnesota 55415

Butterworth & Co., Ltd.
88 Kingsway
London, W.C. 2, England

Cahners Publishing Co., Inc.
221 Columbus Avenue
Boston, Massachusetts 02116

California Council of Civil
Engineers & Land Surveyors
1107 Ninth Street, Suite 811
Sacramento, California 95814

California Legislature
Assembly Committee on Water
Assembly Box 38
State Capitol
Sacramento, California 95814

California. State of. Resources Agency
Sacramento, California 95820

California State Water Resources
Control Board
Department of General Services
Office of Procurement - Stores
Document Section
P.O. Box 20191
Sacramento, California 95820

Callaghan and Co.
165 North Archer Avenue
Mundelein, Illinois 60060

Cambridge University Press
32 East Fifth-seventh Street
New York, New York 10022
(Cambridge, England)

Cast Iron Pipe Research Association
Executive Plaza East
1211 West Twenty-second Street
Oak Brook, Illinois 60521

Cast Iron Soil Pipe Institute
2029 K Street N.W.
Washington, D.C. 20006

Center for Agricultural Publishing
and Documentation
P.O. Box 4
Wageningen, The Netherlands

Center for California Public Affairs
Claremont, California 91711

Center for the Study of Federalism.
See Temple University

Center for Urban Studies. See
University of Chicago

Central Public Health
Engineering Research Institute
Nagpur, India

Certain-teed Products Corp.,
Pipe Division
Ambler, Pennsylvania 19002

Chemical Engineering Magazine
330 West Forty-second Street
New York, New York 10036

Chemical Publishing Co., Inc.
200 Park Avenue, South
New York, New York 10003

Chemical Rubber Co.
18901 Cranwood Parkway
Cleveland, Ohio 44128

Chlorine Institute, Inc.
342 Madison Avenue
New York, New York 10017

Clearinghouse for Federal Scientific
and Technical Information
5285 Port Royal Road
Springfield, Virginia 22151

Clemson University
Environmental Systems Engineering
Department
Clemson, South Carolina 29631

Clemson University
Office of Industrial and Municipal
Relations
112 Riggs Hall
Clemson, South Carolina 29631

Clinton Industries, Inc.
Publishing Division
P.O. Box 1208
32880 Dequindre Avenue
Warren, Michigan 48092

J.B. Clow & Sons, Inc.
Cast Iron & Foundry Division
201-299 North Talman Avenue
Chicago, Illinois 60680

Co-Libri
P.O. Box 482
The Hague 2076, Netherlands

College Science Publishers
P.O. Box 798
State College, Pennsylvania 16801

Colorado State University
Engineering Research Center
Fort Collins, Colorado 80521

Colt Industries
Pump Division
3601 Kansas Avenue
Kansas City, Kansas 66106

Columbia University Press
440 West 110th Street
New York, New York 10025

Commonwealth Government Printing
Office
Canberra, New South Wales
Australia

Communications Research
Machines, Inc.
Del Mar, California 92014

Conservation Foundation
1717 Massachusetts Avenue N.W.
Washington, D.C. 20036

Cornell University
Agricultural Waste Management
Program
Riley-Robb Hall
Ithaca, New York 14850

Cornell University Press
124 Roberts Place
Ithaca, New York 14850

Council of Europe
[U.S. distributor] Manhattan
    Publishing Co.
225 Lafayette Street
New York, New York 10012

Council of Planning Librarians
P.O. Box 229
Monticello, Illinois 61856

Coward, McCann & Geoghegan, Inc.
200 Madison Avenue
New York, New York 10016

Crane, Russak & Co.
347 Madison Avenue
New York, New York 10017

Crowell, Collier & Macmillan, Inc.
866 Third Avenue
New York, New York 10022

Dairy and Food Industries Supply
    Association, Inc.
5530 Wisconsin Avenue N.W.
Washington, D.C. 20015

F.A. Davis Co.
1915 Arch Street
Philadelphia, Pennsylvania 19103

Walter de Gruyter, Inc.
162 Fifth Avenue
New York, New York 10010

Marcel Dekker, Inc.
95 Madison Avenue
New York, New York 10016

Denver Regional Council of
    Governments
1776 South Jackson
Denver, Colorado 80210

Directory of Environmental Consultants
P.O. Box 8002
St. Louis, Missouri 63108

Directory Press
P.O. Box 8002
St. Louis, Missouri 63108

Doubleday & Co., Inc.
501 Franklin Avenue
Garden City, New York 11530

Dover Publications, Inc.
180 Varick Street
New York, New York 10014

Drexel University
Dean of Continuing & Cooperative
    Education
Philadelphia, Pennsylvania 19104

Duke University
Department of Civil Engineering
Durham, North Carolina 27706

Dun-Donnelley Publications
466 Lexington Avenue
New York, New York 10017

Dunellen Publishing Co., Inc.
386 Park Avenue, South
New York, New York 10016

Ecology Center Confederation
P.O. Box 1100
Berkeley, California 93301

Educational Publishers, Inc.
1125 North State Parkway
Chicago, Illinois 60610

Elsevier/North Holland
335 Jan Van Galenstraat
P.O. Box 211
Amsterdam-W, Netherlands

Engineers Joint Council
Department P
345 East Forty-seventh Street
New York, New York 10017

Environmental Research and
    Applications, Inc.
24 Danbury Road
Wilton, Connecticut 06897

Environmental Science Services Corp.
Stamford, Connecticut 06904

Publishers

Environment Information Center, Inc.
Film References Department
124 East Thirty-ninth Street
New York, New York   10016

Exposition Press, Inc.
50 Jericho Turnpike
Jericho, New York   11753

Federal Water Pollution Control
    Administration
4676 Columbia Parkway
Cincinnati, Ohio   45226

The Fertilizer Institute
1015 Eighteenth Street N.W.
Washington, D.C.   20036

Field Studies Council
Pembroke, Wales
Great Britain

Fisher Controls Co.
205 South Center
Marshalltown, Iowa   50158

Fleet Press Corp.
156 Fifth Avenue
New York, New York   10010

W.H. Freeman and Co.
660 Market Street
San Francisco, California   94104

Frost & Sullivan, Inc.
106 Fulton Street
New York, New York   10038

Geological Survey of Alabama
Library
P.O. Box O
University, Alabama   35486

George Washington University
PTC Research Institute
Washington, D.C.   20006

Gordon and Breach
440 Park Avenue, South
New York, New York   10016

Willard Grant Press, Inc.
53 State Street
Boston, Massachusetts   02109

Graphic Arts Center
Portland, Oregon   97200

Graphics Management Corp.
1101 Sixteenth Street N.W.
Washington, D.C.   20036

Stephen Greene Press
Box 1000
Brattleboro, Vermont   05301

Grossman Publishers, Inc.
44 West Fifty-sixth Street
New York, New York   10019

Grozier Publishing, Inc.
Warren Avenue
Harvard, Massachusetts   01451

Halsted Press
605 Third Avenue
New York, New York   10016

Harper and Row Publishers
49 East Thirty-third Street
New York, New York   10016

Harvard University Press
79 Garden Street
Cambridge, Massachusetts   02138

Headley Brothers, Ltd.
Invicta Press
Ashford, Kent, England

D.C. Heath & Co.
125 Spring Street
Lexington, Massachusetts   02173

W. Heffer & Sons, Ltd.
Cambridge, England

Her Imperial Majesty's Stationery Office
Department of Health and Social
    Security, Welsh Office
Ministry of Housing and Local
    Government
London, England

Her Imperial Majesty's Stationery
   Office
[U.S. distributor] Pendragon House, Inc.
899 Broadway Avenue
Redwood City, California   94063

Hertillon Press
P.O. Box 424
North East, Maryland   21901

Holt, Rinehart and Winston, Inc.
383 Madison Avenue
New York, New York   10017

Houghton Mifflin Co.
2 Park Street
Boston, Massachusetts   02107

Humanities Press
303 Park Avenue, South
New York, New York   10010

Hydraulic Institute
122 East Forty-second Street
New York, New York   10017

IFI/Plenum Corp.
227 West Seventeenth Street
New York, New York   10011

IIT Research Institute
10 West Thirty-fifth Street
Chicago, Illinois   60616

Illinois. State of. Department of
   Registration and Education
Urbana, Illinois   61801

Illinois State Geological Survey
National Resources Building
Urbana, Illinois   61801

Illinois State Water Survey
Box 232
Urbana, Illinois   61801

Indiana University
Water Resources Research Center
Bloomington, Indiana   47401

Indiana University Press
Tenth and Morton Streets
Bloomington, Indiana   47401

Industrial Equipment Manufacturers
   Council
410 North Michigan Avenue
Chicago, Illinois   60611

Industrial Press, Inc.
200 Madison Avenue
New York, New York   10016

Institute for Research on Human
   Resources. See Pennsylvania
   State University

Institute for Scientific Information
Publications Division
325 Chestnut Street
Philadelphia, Pennsylvania   19106

Institute of Food Technologists
221 North LaSalle Street
Chicago, Illinois   60601

Institute of Paper Chemistry
P.O. Box 1048
Appleton, Wisconsin   54911

Institute of Water Pollution Control
Ledson House
53 London Road
Maidstone, Kent, England

The Institution of Mechanical
   Engineers
1 Birdcage Walk
London, SW1H 9JJ, England

Institution of Water Engineers
6-8 Sackville Street
Piccadilly
London, W1X IDD, England

Instrument Society of America
400 Stanwix Street
Pittsburgh, Pennsylvania   15222

Publishers

Interdevelopment, Inc.
Suite 814
Millard Fillmore Building
2341 South Jefferson Davis Highway
Arlington, Virginia   22202

International Association for Pollution
Control
Suite 303, 4733 Bethesda Avenue N.W.
Washington, D.C.   20014

International Atomic Energy Agency
[U.S. distributor] UNIPUB, Inc.
P.O. Box 433
New York, New York   10016

International City Management
Association
1140 Connecticut Avenue N.W.
Washington, D.C.   20036

International Textbook Co.
Scranton, Pennsylvania   18515

International Underground Water
Institute
Attn: William V. Karr
2270 Bryden Road
Columbus, Ohio   43209

International Wrought Copper Council
6 Bathurst Street
London, W2, England

Interscience Publishers
Division of John Wiley & Sons, Inc.
605 Third Avenue
New York, New York   10016

Interstate Printers and Publishers, Inc.
19-27 North Jackson Street
Danville, Illinois   61832

Intext Educational Publishers
Oak Street and Pawnee Avenue
Scranton, Pennsylvania   18515

Iowa State University Press
Ames, Iowa   50010

Johns Hopkins University Press
Baltimore, Maryland   21218

Johns-Manville Corp.
22 East Fortieth Street
New York, New York   10016

Alfred A. Knopf, Inc.
201 East Fiftieth Street
New York, New York   10022

Robert E. Krieger Publishing Co., Inc.
Box 542
Huntington, New York   11743

LaMotte Chemical Products Co.
Chestertown, Maryland   21620

Lancaster Press, Inc.
Lancaster, Pennsylvania   17604

League of Women Voters
1730 M Street N.W.
Washington, D.C.   20036

Learning Systems Co.
1818 Ridge Road
Homewood, Illinois   60430

Leupold & Stevens, Inc.
P.O. Box 688
600 Meadow Drive N.W.
Beaverton, Oregon   97005

Lewis and Clark College
10015 Terwilliger Boulevard S.W.
Portland, Oregon   97219

Lexington Data, Inc.
Box 311
Lexington, Massachusetts   02173

J.B. Lippincott Co.
East Washington Square
Philadelphia, Pennsylvania   19105

Liverpool University Press
123 Grove Street
Liverpool 7, England

Louisiana Polytechnic Institute
Animal Foundation
Ruston, Louisiana   71271

Louisiana State University
Louisiana Water Resources Research
Institute
[Available from L.S.U. Bookstore]
Baton Rouge, Louisiana 70803

McGraw-Hill Book Co.
1221 Avenue of Americas
New York, New York 10020

McGraw-Hill Information Systems Co.
330 West Forty-second Street
New York, New York 10036

David McKay Co., Inc.
750 Third Avenue
New York, New York 10017

McLoughlin Research Associates, Inc.
10516 Estate Lane
Dallas, Texas 75238

The Macmillan Co.
886 Third Avenue
New York, New York 10022

Macmillan Journals, Ltd.
Brunel Road
Basingstoke, Hants, England

Management Sourcebook, Inc.
347 Madison Avenue
New York, New York 10017

Manufacturing Chemists Association
1825 Connecticut Avenue N.W.
Washington, D.C. 20009

Marine Science Institute. See
University of California

Maryland Division of Economic
Development
State Office Building
Annapolis, Maryland 21401

Memphis State University Press
Memphis, Tennessee 38111

Merck and Co., Inc.
126 East Lincoln Avenue
Rahway, New Jersey 07065

Charles B. Merrill Publishing Co.
Columbus, Ohio 43216

Michigan State University
East Lansing, Michigan 48823

Midwest Environmental
Management, Inc.
P.O. Box 82
Maumee, Ohio 43537

Mills & Boon, Ltd.
London, England

Missouri Water Well & Pump
Contractors Association, Inc.
Box 250
Rolla, Missouri 65401

The MIT Press
28 Carleton Street
Cambridge, Massachusetts 02142

C.V. Mosby Co.
11830 Westline Industrial Drive
St. Louis, Missouri 63141

Municipal Finance Officers Association
1313 East Sixtieth Street
Chicago, Illinois 60637

National Academy of Sciences
2101 Constitution Avenue
Washington, D.C. 20418

National Agricultural Chemicals
Association
1155 Fifteenth Street N.W.
Washington, D.C. 20005

National Association of Corrosion
Engineers
2400 West Loop, South
Houston, Texas 77027

National Association of Manufacturers
277 Park Avenue
New York, New York 10017

Publishers

National Association of Regulatory
Utility Commissioners
1102 I.C.C. Building
P.O. Box 684
Washington, D.C. 20004

National Association of Secondary
Materials Industries, Inc.
330 Madison Avenue
New York, New York 10017

National Council on Radiation
Protection and Measurements
4201 Connecticut Avenue N.W.
Washington, D.C. 20008

National Field Investigations Center
555 Ridge Avenue
Cincinnati, Ohio 45213

National Fire Protection Association
60 Batterymarch Street
Boston, Massachusetts 02110

National Foundation for Environmental
Control, Inc.
151 Tremont Street
Boston, Massachusetts 02111

National Institute of Municipal Law
Officers
839 Seventeenth Street N.W.
Washington, D.C. 20006

National Lime Association
5010 Wisconsin Avenue N.W.
Washington, D.C. 20016

National Pollution Control Conference
and Exposition
P.O. Box 22372
Houston, Texas 77027

National Press
850 Hansen Way
Palo Alto, California 94304

National Safety Council
425 North Michigan Avenue
Chicago, Illinois 60611

National Sanitation Foundation. See
University of Michigan

National Service to Regional Councils
1700 K Street N.W.
Washington, D.C. 20006

National Swimming Pool Institute
2000 K Street N.W.
Washington, D.C. 20006

National Technical Information
Service
Department of Commerce
Springfield, Virginia 22151

National Water Institute
744 Broad Street
Newark, New Jersey 07102

National Watershed Congress
1025 Vermont Avenue N.W.
Washington, D.C. 20005

The Natural History Press
Garden City, New York 11530

New Jersey Department of Agriculture
P.O. Box 1888
Trenton, New Jersey 08625

New York State Department of
Environmental Conservation
Office of Recovery, Recycling and
Reuse
Albany, New York 12206

New York State Department of Health
845 Central Avenue
Albany, New York 12206

North American International
P.O. Box 28278
Central Station
Washington, D.C. 20005

North Carolina State University
Industrial Extension Service
P.O. Box 5506
Raleigh, North Carolina 27607

North Carolina State University
Water Resources Research Institute
124 Riddick Building
Raleigh, North Carolina   27607

Northern Publishers
Aberdeen, Scotland

North Holland Publishers
Amsterdam, Holland

W.W. Norton and Co., Inc.
55 Fifth Avenue
New York, New York   10003

Noyes Data Corp.
118 Mill Road
Park Ridge, New Jersey   07656

Ocean Engineering Information
    Service
P.O. Box 989
La Jolla, California   92037

Oklahoma State University
Center for Water Research in
    Engineering
Bioenvironmental Engineering
School of Civil Engineering
Stillwater, Oklahoma   74074

Oklahoma State University Press
Stillwater, Oklahoma   74074

Oligodynamics Press
P.O. Box 35774
Houston, Texas   77035

Optosonic Press
Box 883
Ansonia Station
New York, New York   10023

Organization for Economic
    Co-Operation and Development
2 rue Andre-Pascal
75 Paris 16e, France

Oriel Press, Ltd.
Newcastle upon Tyne, England

Oxford University Press
200 Madison Avenue
New York, New York   10016

Pegasus
4300 West Sixty-second Street
Indianapolis, Indiana   46226

Pegasus Press
350 Third Avenue
New York, New York   10022

Pemberton Press (Jenkins Publishing Co.)
Box 2085
Auxtin, Texas   78767

Pendragon House, Inc. See Her
    Imperial Majesty's Stationery
    Office

Pennsylvania State University
Institute for Research on Human
    Resources
University Park, Pennsylvania   16802

The Pennsylvania State University Press
215 Wagner Building
University Park, Pennsylvania   16802

Peter Peregrinus Ltd.
P.O. Box 8
Southgate House
Stevenage, Herts, SG1 1HQ, England

Pergamon Press, Inc.
Maxwell House
Fairview Park, Elmsford, New York
    10523

Plenum Publishing Corp.
227 West Seventeenth Street
New York, New York   10011

Pocket Books, A Division of Simon &
    Schuster
630 Fifth Avenue
New York, New York   10020

Portland Cement Association
Old Orchard Road
Skokie, Illinois   60076

Publishers

Praeger Publishers
111 Fourth Avenue
New York, New York   10003

Premier Press
P.O. Box 4428
Berkeley, California   94704

Prentice-Hall, Inc.
Engelwood Cliffs, New Jersey   07632

Princeton University Press
Princeton, New Jersey   08540

Printers' Corner
18 Shantisadan Society
Ellisbridge
Ahmedabad-6, India

Priority Press
London Road
St. Albans, Herts, England

PTC Research Institute. See George
   Washington University

Public Administration Service
1313 East Sixtieth Street
Chicago, Illinois   60637

Public Personnel Association
1313 East Sixtieth Street
Chicago, Illinois   60637

Public Service Research, Inc.
7050 S.W. Eighty-sixth Avenue
Miami, Florida   33143

Public Works Publications
200 South Broad Street
Ridgewood, New Jersey   07451

Purdue University
Office of University Editor
Building D, South Campus Courts
Lafayette, Indiana   47907

Reinhold Publishing Corp.
600 Summer Street
Stamford, Connecticut   06904

Resources for the Future, Inc.
1755 Massachusetts Avenue N.W.
Washington, D.C.   20036

Rodale Press, Inc.
33 East Minor Street
Emmaus, Pennsylvania   18049

The Ronald Press Co.
79 Madison Avenue
New York, New York   10016

Royal Swedish Academy of Sciences
Universitetsforlaget
P.O. Box 307
Blindern, Oslo 3, Norway

Rural Water Commission
Box DM
221 North LaSalle Street
Chicago, Illinois   60601

Rutgers University
Center for Urban Policy
Research and Conferences Department
University Extension Division
New Brunswick, New Jersey   08903

Rutgers University
Water Resources Research Institute
New Brunswick, New Jersey   08903

Sage Hill Publishers, Inc.
435 Hudson Street
New York, New York   10014

W.B. Saunders Co.
218 West Washington Square
Philadelphia, Pennsylvania   19105

Scott, Foresman & Co.
1900 East Lake Avenue
Glenview, Illinois   60025

Scranton Publishing Co.
434 South Wabash Avenue
Chicago, Illinois   60605

Skeist Laboratories, Inc.
Livingston, New Jersey   07039

Publishers

Snyder Oceanography Services
P.O. Box 98
Jupiter, Florida 33458

Society of Chemical Industry
14 Belgrave Square
London, S.W.1, England

Soil Conservation Society of America
7515 Ankeny Road N.E.
Ankeny, Iowa 50021

Soil Science Society of America, Inc.
677 South Segoe Road
Madison, Wisconsin 53711

Spartan Books
Hayden Book Co., Inc.
50 Essex Street
Rochelle Park, New Jersey 07662

Special Libraries' Association
235 Park Avenue, South
New York, New York 10003

Springer-Verlag New York, Inc.
175 Fifth Avenue
New York, New York 10010

Stanford Research Council
Editorial Service
Menlo Park, California 94025

Steel Structures Painting Council
4400 Fifth Avenue
Pittsburgh, Pennsylvania 15213

Swan House Publishing Co.
P.O. Box 638
Binghamton, New York 13902

Swedish Natural Research Council,
Editorial Service
Box 23 136, S-104
35 Stockholm, Sweden

Syracuse University Press
Box 8, University Station
Syracuse, New York 13210

Technical and Economical
Publishing Corp.
P.O. Box 8893
440 Meppelweg
The Hague, Netherlands

Technical Guidance Center for
Industrial Environmental Control.
See University of Massachusetts

Technomic Publishing Co., Inc.
265 West State Street
Westport, Connecticut 06880

Temple Press Books
London, England

Temple University
Center for the Study of Federalism
Philadelphia, Pennsylvania 19122

Texas. State Bar of
Austin, Texas 78711

Texas Water Utilities Association
2202 Indian Trail
Austin, Texas 78703

Charles C. Thomas, Publisher
301-327 East Lawrence Avenue
Springfield, Illinois 62717

Thomas Printing & Publishing
724 Desnoyer Street
Kaukauna, Wisconsin 54130

UNESCO
National Agency for International
Publication
317 East Thirty-fourth Street
New York, New York 10016

UNESCO
[U.S. distributor] UNIPUB, Inc.
P.O. Box 433
New York, New York 10016

United Nations Publications
United Nations, Room LX 2300
New York, New York 10017

Publishers

University of Arizona Press
P.O. Box 3398
College Station
Tucson, Arizona 85700

University of California
Marine Science Institute
Oil Spill Information Center
Santa Barbara, California 93106

University of California Press
2223 Fulton Street
Berkeley, California 94720

University of Chicago
Center for Urban Studies
1307 East Sixtieth Street
Chicago, Illinois 60637

University of Chicago
Department of Geography
5828 South University Avenue
Chicago, Illinois 60637

University of Chicago Press
5801 Ellis Avenue
Chicago, Illinois 60637

University of Delaware
Publications Office
College of Marine Studies
Newark, Delaware 19711

University of Illinois
American Association of Professors
in Sanitary Engineering
3230 Civil Engineering Building
Urbana, Illinois 61801

University of Illinois
Bureau of Economic and Business
Research
Urbana, Illinois 61801

University of Illinois
Department of Civil Engineering
Engineering Publications Office
112 Engineering Hall
Urbana, Illinois 61801

University of Illinois Bulletin
114 Altgeld Hall
Urbana, Illinois 61801

University of Massachusetts
Technical Guidance Center for
Industrial Environmental Control
Amherst, Massachusetts 01002

University of Miami Press
Drawer 9088
Coral Gables, Florida 33124

University of Michigan
College of Engineering
Ann Arbor, Michigan 49104

University of Michigan
National Sanitation Foundation
P.O. Box 1468
School of Public Health
Ann Arbor, Michigan 48106

University of Michigan Press
615 East University
Ann Arbor, Michigan 48106

University of Minnesota Press
2037 University Avenue S.E.
Minneapolis, Minnesota 55455

University of Missouri Publications
B-9 Whittern Hall
Columbia, Missouri 65201

University of Montana
School of Forestry
Missoula, Montana 59801

University of New Mexico Press
Albuquerque, New Mexico 87106

University of North Carolina
Department of Environmental Sciences
and Engineering
Box 630
Chapel Hill, North Carolina 27514

University of Northern Iowa
Iowa Academy of Science
Cedar Falls, Iowa 50613

University of Rhode Island
Division of Engineering Research
108 Bliss Hall
Kingston, Rhode Island   02881

University of Southern California
School of Engineering
Los Angeles, California   90007

University of Texas
Association of Environmental
    Engineering Professors
305 Engineering Labs Building
Austin, Texas   78712

University of Texas
Center for Research in Water Resources
Austin, Texas   78757

University of Texas Press
P.O. Box 7819
Austin, Texas   78712

University of Toronto Press
Front Campus
University of Toronto
Toronto 5, Canada

University of Washington Press
Seattle, Washington   98105

University of Wisconsin
Cooperative Extension Programs
Madison, Wisconsin   53703

University of Wisconsin
Department of Law, University
    Extension
Madison, Wisconsin   53706

University of Wisconsin
Water Resources Center
Hydraulic and Sanitary Laboratory
Madison, Wisconsin   53706

University of Wisconsin Press
P.O. Box 1379
Madison, Wisconsin   53701

The Urban Institute
2100 M Street N.W.
Washington, D.C.   20037

UOP - Johnson Division
University Oil Products Co.
Box 3118
St. Paul, Minnesota   55165

U.S. and World Publishing, Inc.
225 Lafayette Street
New York, New York   10012

U.S. Army Corps of Engineers
    Waterways Experiment Station
P.O. Box 631
Vicksburg, Mississippi   39180

U.S. Army Engineer Division
North Central
536 South Clark Street
Chicago, Illinois   60605

U.S. Borax
3075 Wilshire Boulevard
Los Angeles, California   90075

U.S. Department of Commerce
Clearinghouse
Springfield, Virginia   22151

U.S. Department of Commerce
National Industrial Pollution
    Control Council
Washington, D.C.   20230

U.S. Department of the Interior
Bureau of Mines
Publication Section
4800 Forbes Avenue
Pittsburgh, Pennsylvania   15213

U.S. Department of the Interior
Geological Survey
Washington, D.C.   20242

U.S. Department of the Interior
Office of Saline Water
Washington, D.C.   20240

U.S. Environmental Protection Agency
4676 Columbia Parkway
Cincinnati, Ohio   45206
(See also Advanced Waste Treatment
Research Laboratory and Analytical
Quality Control Laboratory)

Publishers

U.S. Government Printing Office
Washington, D.C. 20402

U.S. Public Health Service
Office of Public Information
Ballston Center Tower 2
801 North Randolph
Arlington, Virginia 22203

Vanderbilt University
Department of Environmental and
   Water Resources Engineering
Box 133, Station B
Nashville, Tennessee 37203

Vanderbilt University Press
Nashville, Tennessee 37205

D. Van Nostrand Co.
450 West Thirty-third Street
New York, New York 10001

Van Nostrand Reinhold Co.
450 West Thirty-third Street
New York, New York 10001

Vocational Guidance Manuals
233 East Forty-fifth Street
New York, New York 10017

Wadsworth Publishing Co., Inc.
10 Davis Drive
Belmont, California 94002

Walker and Co.
720 Fifth Avenue
New York, New York 10019

The Wall Street Journal
30 Broad Street
New York, New York 10004

Washington Square Press, Inc.
630 Fifth Avenue
New York, New York 10020

Washington State University
Technical Extension Services
Pullman, Washington 99163

Water Information Center
Water Research Building
Manhasset Isle
Port Washington, New York 11050

Water Pollution Control Federation
3900 Wisconsin Avenue
Washington, D.C. 20016

Water Resources Publications
1912 Sequoia Street
Fort Collins, Colorado 80521

Wesleyan University
Industrial Waste Laboratory
Middletown, Connecticut 06457

Western Michigan University
Institute of Public Affairs
Kalamazoo, Michigan 49001

John Wiley & Sons, Inc.
605 Third Avenue
New York, New York 10016

Williams and Wilkins Co.
428 East Preston Street
Baltimore, Maryland 21202

Winchester Press
460 Park Avenue
New York, New York 10022

World Health Organization
[U.S. distributor] American Public
   Health Association, Inc.
1015 18th Street N.W.
Washington, D.C. 20036

World Meteorological Organization
[U.S. distributor] UNIPUB, Inc.
P.O. Box 433
New York, New York 10016

Yale University Press
92 A. Yale Station
New Haven, Connecticut 06520

York Press Publishers
101 East Thirty-second Street
Baltimore, Maryland 21218

INDEXES

AUTHOR INDEX

A

Ahlberg, N.R.   506

Alberi, L.T.   111

American City Magazine   613

American Society of Civil Engineers
204, 207, 307, 366

Antonie, R.L.   319

Argaman, Y.   203

Associated Water and Air Resources
Engineers, Inc.   376

Atkins, P.F.   338a

Austin, Texas, City of   311

B

Babbitt, H.E.   a01

Bacon, V.W.   344c

Baier, D.C.   412

Balakrishnan, S.   516

Baldwin, R.B.   801

Banker, R.F.   608

Barsom, G.   324

Bartch, E.H.   321

Barth, E.F.   105

Baumann, R.E.   a01

Beebe, R.L.   345

Bell, G.R.   341a, 344b

Bella, D.A.   109

Benarde, M.A.   418

Bendixen, T.W.   402, 428b

Berg, G.   368, 372a

Bishop, D.F.   328

Black & Veatch, Consulting Engineers
339

Blecker, H.G.   606

Bloom, R.S.   331

Bolton, R.L.   a02

Bonner, W.F.   330

Bouwer, H.   360, 411

Boyko, B.I.   506

Boyle, W.   302

Author Index

Author Index

Author Index

Author Index

TITLE INDEX

A

Advanced Wastewater Treatment   a06

Advanced Wastewater Treatment Using Powdered Activated Carbon in Recirculating Slurry Contractor Clarifiers   345

Advances in Water Pollution Research: Proceedings of the 5th International Conference   201, 372b

Advances in Water Quality Improvement   323, 327, 338j

Aeration in Wastewater Treatment   316

American Sewerage Practice, Vol. 1: Design of Sewers   a18

American Sewerage Practice, Vol. 3: Disposal of Sewage   a19

Anaerobic Sludge Digestion   509

Applied Stream Sanitation   a31

Aquatic Chemistry   a30

B

Biological Aspects of Thermal Pollution   422c

C

Calcium Phosphate Precipitation in Wastewater Treatment   341f

Capital and Operating Costs of Pollution Control Equipment Modules   606

Chemistry for Sanitary Engineers   a29

Combined Sewer Separation Using Pressure Sewers   207

Cost-Effectiveness: The Economic Evaluation of Engineered Systems   603

Costs of Water Pollution Control, Proceedings of a National Symposium   108

Cost to the Consumer for Collection and Treatment of Wastewater   610

D

Desalination by Reverse Osmosis   350

Design and Contruction of Sanitary and Storm Sewers   204

Design Guides for Biological Wastewater Treatment Processes   311

Disposal of Sewage and Other Waterborne Wastes   a15

Title Index

SUBJECT INDEX

References are to entry numbers, with the following exceptions. Since entries from Appendix A may also appear in the text, such textual references are followed by a page number in parentheses: a17(p.27); references to Appendices B-F are indicated by page number only: p.140.

A

Acid spoil banks 528

Activated sludge: description of process 313, 341b, 341g, a07, a17(p.27)
design considerations 313, 314, 315, 317, a12, a17(p.27)
operation and maintenance 313, 317
process modifications 310, 317, a17 (p.27)
thickening 310, 314

Adsorption. See Carbon adsorption

Advanced wastewater treatment (AWT) 105, 328, 332, 333, 611, a06, a21. See Dissolved organics removal; Dissolved solids removal; Nitrogen removal; Phosphorus removal; Suspended solids removal

Aerated grit chambers 308

Aerated lagoons. See Lagoons, aerated

Aeration 315, 316, 322, 323

Aerators 316, 322, 323

Aerobic sludge digestion 506, 507, 513. See Sludge, digestion

Aerosols 421

Agricultural Research Service 519

Agricultural reuse 402-405, 421. See Effluent, reuse

Agriculture. See Crops; Cropland

Air pollution 517, a04, a27, a28, p.133, p.135, p.139, p.140; laws 112, p.140

Air stripping of nitrogen 337, 338, 341

Algae 320, 324, 336, 337, 409, 511, 708

Alum 330, 340

Aluminum salts 339, 340

Ammonia nitrogen removal 334, 338, 338a, 338d, 338i

Ammonia removal 328

Subject Index

installations 428; Whittier Narrows 361

Campbell Soup Company 358

Carbon adsorption 328, 329, 330, 331, 345, 347, 349, a06(p.40), a12(p.37), a32(p.48)

Carbon regeneration 330, a06

Catfish farming 410, 529

Centrifugation 513, 514, 515

Chemical coagulation 341a, 372b, a06. See Coagulation

Chemical oxygen demand (COD) 345

Chemical precipitation. See Precipitation, chemical

Chemical stabilization 512

Chemical treatment. See Biological-chemical treatment; Physical-chemical treatment

Chemical unit processes 305, a11 (p.26), a21(p.26), a25(p.27), a26, a32(p.28)

Chemistry a29, a30, p.145; aquatic a11(p.26)

Chicago Prairie Plan 527

Chlorination: application 338, 363, 364, 365, 366, 367, 372e, 413, a11(p.55), a21(p.57) breakpoint 337, 363, 364, a28 (p.57) history 364 theory 364, a11(p.55), a21(p.57), a29(p.57), a32(p.56)

Chlorinators 364, 366

Chlorine: chemistry 363, 364, a11 (p.55), a29(p.57), a32(p.56); con-

tact chambers 365, a11(p.55), a21 (p.57); residual 365, 366, a29(p.57)

Cities. See Municipalities and for individual cities, see under state name

Civil engineering p.146

Clarification. See Sedimentation

Clarification-adsorption 331c

Clarifiers 345

Climatic effects on irrigation 402

Clinoptilolite 338i

Coagulation 341a, a06, a32

Coagulation-precipitation 331

COD. See Chemical oxygen demand

Code of Federal Regulations p.140

Coliform. See Bacteria

Collection systems 103, 104, 203-208

Colloid chemistry. See Chemistry

Combined sewer overflows 201, 209, 210, 214, 301

Combined sewers. See Sewers, combined; Regionalization

Compaction. See Thickening

Composting 519, 525

Computer applications for cost analyses 606

Conditioning of sludge 502, 504, 513, 515, a21(p.80)

Construction planning a09

Contact stabilization 315

Subject Index

Cooling devices 422

Cost: comparison 107, 207, 342, 348, 413, 423, 518, 604, 612
consumer 613
data 107, 339, 345, 354, 410, 428, 501, 516, 525, 527, 530, 606-611, a09(p.6)
escalation 602, 609
estimates 346, 514, 608, a32
indexing 602
operating 212, 214, 358, 607, 608, 609

Cost-effectiveness 108, 603, 604, a09

Cropland, disposal on 523, 524

Crops: irrigation 106, 403, 405, 406, 407; response to wastewater 356, 358, 359, 421; yields 407, 524

D

Delaware River Basin 703

Demineralization 332, 333, 348, 349

Denitrification 335-337, 338e, 338g, 338h, 341, 360, a06(p.40), a21 (p.41)

Desalination 350, p.140

Detergents p.141

Dewatering of sludge. See Sludge, dewatering

Digestion. See Sludge, digestion

Disease 418, 421, 518. See Epidemiology

Disinfection 362-372, 372a-372f, a11(p.55), a21(p.57), a29(p.57), a32(p.56)

Dissolved minerals 404

Dissolved organics removal a32(p.48). See also Carbon adsorption; Ozonation; Reverse osmosis

Dissolved solids removal 348-350, 350a-350c, a32(p.48). See also Electrodialysis; Ion exchange; Reverse osmosis

Drying beds 515

E

Ecology 102, 109, 707, 708, a03, a04, a08

Economic considerations 108, 601, 602, 604. See also Cost; Cost-effectiveness; Planning, economic

Effluent: disposal 352, 356, 422-428, 422a-422c, 428a-428c, a16(p.73, p.75), a21(p.73), a24(p.73), a31 (p.74)
filtration 344, a06, a21(p.41)
reuse 330, 401-421, a12, a12(p.66)
treatment p.133 (see also Wastewater, treatment)

Electrochemical treatment 337

Electrodialysis 350c, a21(p.41), a32 (p.49)

Electromagnetic energy a04

Electron irradiation, accelerated 371

Electrons, high-energy 371

Engineering economics 603, 606, 608

Engineering News Record Construction Cost Index 338

Environment 706, 707, a04, p.133, p.134, p.140, p.145
engineering a07, a08, a27, a28, p.140
man and 101, 521, 709, 710, p.146
planning 101, 109, 111, 428b, 708, 709, a04, a28, p.140, p.141, p.146

Subject Index

O

Ocean. See Streams

Ocean disposal 423, 425, 520, 521, 522

Odors 503, 510

Ohio: Miami River Valley 703; Napoleon 358

Oklahoma: Ada 359; Oklahoma City 408

Ontario: sludge treatment 506; Toronto, sewer system 211

Operation of treatment plants. See Wastewater treatment plant, operation

Ordinances. See Legislation, pollution control

Oregon, environmental clean up in 706

Outfall, submarine 423, 424, 520, 521

Overland flow treatment 352, 353, 354, 358-359. See also Groundwater; Irrigation

Oxidation ponds. Lee Lagoons, oxidation

Oxygen: activated sludge 317; demand a21(p.73), a24(p.73); sag a16(p.73), a24(p.73), a31; transfer 316, a11(p.26), a21(p.26)

Ozanation 371, 372, 372c, a06 (p.40), a21

Ozone 364, 370, 372, 372d

P

Parasite transmission 529

Pathogens 418, 420, 421, 523, a22. See Bacteria; Virus

PCB. See Polychlorinated biphenyl

Pennsylvania, sludge treatment in 507

Pennsylvania State University 356

Percolation 426, a16(p.75)

Percolation tests 312, 351. See Infiltration-percolation

Pesticides 401, p.134

Phosphates, precipitation of 341c, 341d

Phosphorus 303, 502, a30

Phosphorus removal 320, 332, 333, 338j, 339-341, 341a-341i, 359, 403, 612, a21(p.41). See also Advanced wastewater treatment

Physical-chemical treatment 105, 328, 329, 330, 331, 331a-331f, 341, 416, a06(p.40), a11(p.37), a12(p.37), a21(p.37), a32(p.37)

Physical chemistry. See Chemistry

Physical unit operations 305, 306, 307, a21(p.26, p.28), a32(p.26, p.28). See also Biological unit treatment processes; Chemical unit processes; Unit operations

Physiochemical treatment. See Physical-chemical treatment

Pipeline transport of sludge 518

Plankton 708. See also Biota

Planning: economic 604, 605, 611, 613, a09, p.140; environmental 109, 705, 710; regional 612, 614, 703, 704, 705; urban 702, 704; wastewater management 101, 110, 611, a09(p.8)

Subject Index

Subject Index

W

Washington, Seattle 520

Washington, D.C. 328

Waste dilution 701

Wastewater: analyses 376–378, a29
(p.62)
characteristics 103, 201, 301, 302,
304, 311, 355, a11, a11(p.24),
a21(p.24), a23(p.25), a29(p.62)
collection 103, 104, 378, 610,
a01, a11, a20, a21, p.135
disposal 103, 320, 352, 353, 354,
401, 421, 422–428, 428a–428c,
a04, a10, a11, a15, a16, a19,
a20, a21(p.73), a24(p.73), a31,
p.133, p.135
engineering p.136
flowrates 201, 311, a01(p.13), a21
(p.13)
household 302, 304, 359
industrial 331c, 355, 358, 376,
521, a23
management 101, 103, 703, 706,
707, a09, p.135
municipal 328, 329, 330, 331c,
331d, 344b, 344d, 350a, 355,
356, 521
reclamation 333, 350a, 350b, 356,
361, 404, 413, 415, 416, 417,
a12(p.66), p.140
reuse 103, 320, 331f, 333, 401–
409, 411–417, 419, 604
sources 201, 202, 209, 210, 301
treatment 103, 105–107, 210, 307,
311, 317, 610, a01, a02, a05,
a10, a11, a18, a20, a21, a32,
p.133, p.135, p.136

Wastewater treatment plant 373–375,
509, p.136
costs 602, 603, 606, 608, 609, 610
design 307, 331b, 373
maintenance 373, 375
operation 311, 319, 373, 374, 375,
416, a01
planning 374, 605, a09
staffing 373, 374

Wastewater treatment studies 319,
328, a21

Water collection and storage 201,
a10, a11, p.141

Water distribution a10

Water pollution 101, 102, 201, 202,
301, 425, 521, 709, 710, a04, a10,
a11, a16, a22, a24, a27, p.133,
p.135, p.136, p.139. See Streams,
pollution; Water quality

Water pollution control 112, 202,
377, 424, 614, 706, 707, 801,
a05, a07, a12, p.136
cost 108, 614
laws 112, 521, 611, 802, 803,
804, 805 (see Legislation, pollution
control)
regulations 110, 521, 802

Water quality: criteria 415, 417,
422, a12, a16, a24(p.73), p.139;
management 410, 705, 707, a07,
a12, a16, a32, p.136, p.139,
p.145–46

Water resources a04, p.141, p.146,
p.147

Water storage. See Water collection
and storage

Water supply a05, a10, a11, a28,
p.141

Weirs a01(p.29)

Wisconsin: Pewaukee 319; wastewater
disposal 427; Westby 402

Woods Hole Oceanographic Institution
409

Y

Yeast production 530

Yield coefficients, bacterial 306